£35.00

Geotextiles in filtration and drainage

Proceedings of the conference *Geofad '92: Geotextiles in filtration and drainage* organised by the UK Chapter of the International Geotextile Society and held at Churchill College, Cambridge, UK on 23 September 1992

Edited by Stephen Corbet and John King

Thomas Telford, London

Published by Thomas Telford Services Ltd, Thomas Telford House, 1 Heron Quay, London E14 4JD

First published 1993

Distributors for Thomas Telford books are
USA: American Society of Civil Engineers, Publications Sales Department, 345 East 47th Street, New York, NY 10017-2398
Japan: Maruzen Co Ltd, Book Department, 3-10 Nihonbashi 2-chome, Chuo-ku, Tokyo 103
Australia: DA Books and Journals, 11 Station Street, Mitcham 2131, Victoria

A catalogue record for this book is available from the British Library

Classification
Availability: Unrestricted
Content: Collected papers
Status: Established knowledge
User: Civil engineering designer, academics, contractors, geosynthetic manufacturers

ISBN: 0 7277 1924 6

Printed and bound in Great Britain

Preface

Geofad '92 was held at the University of Cambridge on 23 September 1992, when selected authors presented papers on the design, specification, testing, application and developments in the use of geotextiles in filtration and drainage.

The Conference was initiated to provide information to practising engineers on the benefits of the use of geotextiles and to illustrate the design and installation techniques required to provide efficient and reliable drainage and filtration systems.

This book comprises the papers and the resulting lively and stimulating discussion which was most ably directed and encouraged by the Chairman for the Conference, Dr Malcolm Bolton. The information contained in this book will enable the engineer to make effective and confident use of geotextiles in conjunction with natural materials. It will be seen that geotextiles are well developed and understood products which are of great value in enhancing the performance of traditional drainage materials, thus providing improved reliability and cost effectiveness.

We record our gratitude to the sponsoring organisations: Don & Low Ltd, Du Pont Ltd, MMG Civil Engineering Systems Ltd, and the UK Chapter of the International Geotextiles Society. Our thanks also go to the Organising Committee: Stephen Corbet (G. Maunsell & Partners), John King (Don & Low Ltd), Alan Petri (MMG Civil Engineering Systems Ltd) and David Wilson (CIS Construction Services Ltd). We also acknowledge the valuable administration assistance given by G. Maunsell & Partners.

We also thank Professor Kerry Rowe, President of the International Geotextiles Society for his presentation of the Keynote Address and for his contributions in the subsequent proceedings, Dr Malcolm Bolton (University of Cambridge) for his most efficient chairmanship of the Conference, the session reporters John Nixon (Netlon Ltd.), Chris Lawson (Exxon Chemical

Geopolymers) and Rick Woods (University of Surrey). Finally we thank the authors and delegates who contributed to a valuable and enjoyable conference.

Stephen Corbet and John King

Contents

Some challenging applications of geotextiles in filtration and drainage

R. K. ROWE, Geotechnical Research Centre, Faculty of Engineering Science, The University of Western Ontario, London, Canada

Two important applications for geotextiles as filters in situations where there is significant potential for failure are examined. Factors discussed include: the significance of variability in the filtration characteristics of the geotextile and the grainsize distribution of the soil it is to protect; construction specifications and quality control; and the potential for chemical and biological clogging of geotextiles used in leachate collection systems.

1. Introduction

Filtration and drainage represent a major area of application for geotextiles. Traditionally they have been used in conventional civil engineering applications and in agricultural drainage. More recently there has been a substantial use of geotextiles as filters in waste disposal applications, particularly in North America. Geotextiles have proven to be very effective in these applications, and this has been well documented in numerous conference proceedings, including the four International Conferences on Geotextiles and Related Products (Paris 1977, Las Vegas 1982, Vienna 1986, The Hague 1990).

Papers presented at this present conference deal with design and specifications, filter performance and future developments. The design methodology is generally well established and in most applications geotextiles can be very easily and successfully used. Rather than review these situations (which represent the majority of cases), the objective of the present paper is to initiate discussion concerning the use of geotextile filters in two important applications where there is significant potential for failure.

Geotextiles in filtration and drainage. Thomas Telford, London, 1993

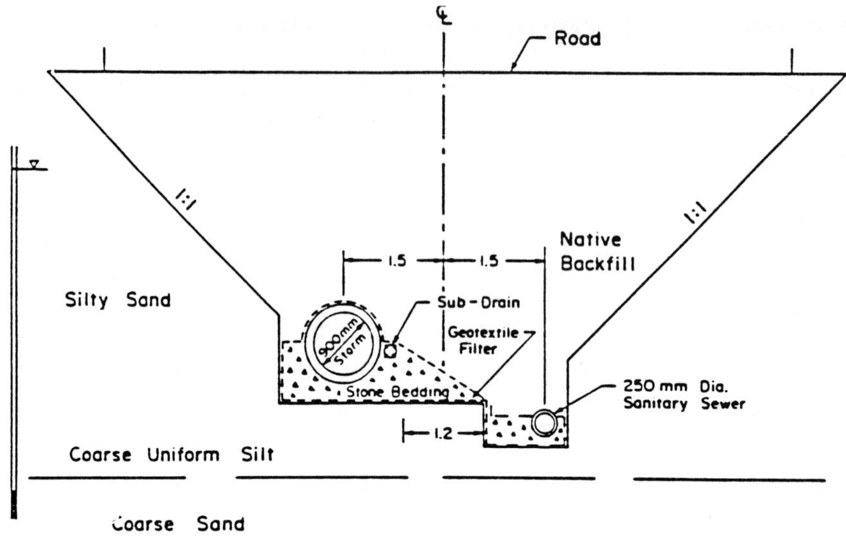

Fig. 1. Geotextile filter for a sewer pipe installation. Potential for failure due to upward gradient from underlying aquifer

2. Geotextile filters where there is potential for piping

In the majority of filtration applications for geotextiles the water flow is either horizontal or has a downward component. However in some applications the hydraulic gradient is upward and hence can induce piping of the soil. Figure 1 shows a schematic of a sewer installation where a geotextile is used as a filter between a silty sand/silt and the stone bedding around the pipes. In this particular example a drainage pipe is also installed to lower the water table below the overlying road, and in so doing induces an inward and upward gradient from an underlying more permeable unit. Even in the absence of the drainage pipe, hydraulic defects in the sewer pipe (e.g. due to an imperfect seal) have the potential for creating a situation where there are upward gradients. In this application the geotextile filter can play a critical role since it must prevent piping of the silt/sand into the coarse stone bedding. In the event that it does not do so, settlement and failure of the sewer is the likely consequence.

Figure 2 shows a detail of a barrier system beneath a landfill. In this particular design the landfill is designed as a 'hydraulic trap' where the leachate level in the landfill is to be maintained at a level below the potentiometric surface in an underlying aquifer. In this case a deliberate attempt is made to induce groundwater flow into the landfill in order to reduce the potential outward diffusive migration of contaminants. In this case the lower geotextile is intended to act as a filter which will allow the movement of water into the 'hydraulic control layer' while preventing any significant movement of soil into the granular hydraulic control layer. (A discussion of the motivation for this type of design is beyond the scope of the present paper and the interested reader is referred to Rowe (1991a, b, c, 1992). Again the geotextile plays a critical role since there is the potential for piping and any significant movement of soil into the hydraulic control layer will reduce the hydraulic conductivity and performance of this layer.

In these two situations it is critical to allow the passage of water without significant movement of fines (and hence without piping). If one accepts that geotextiles can adequately perform this function (which they have) there are three key issues to be considered:

(a) how to design and select a suitable geotextile filter;

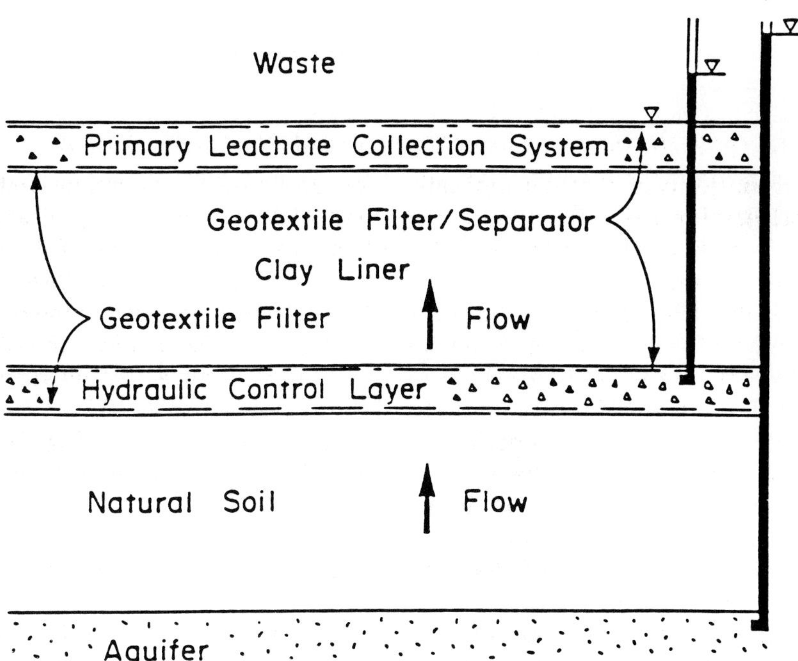

Fig. 2. Geotextile filter for design of a 'hydraulic trap' involving upward movement of water from an aquifer into a granular 'hydraulic control layer'

(b) the level of variability in geotextile properties (notably in the filtration opening size, FOS) that can be tolerated and the relationship between this and the potential variability of the soil itself.
(c) the need for adequate construction specifications and for quality control.

Given data on the grainsize distribution of the soil to be protected, there are a number of design procedures which can be adopted (e.g. Giroud, 1982; Faure, et al., 1986, to list but a few). For example, Faure et al.,1986, recommend that for filtration applications (such as that considered here) the geotextile should be selected such that

$$FOS \leq 1.5d_{85} \tag{1}$$

where FOS is the filtration opening sizeand d_{85} is the point on the soil grading curve at which 85% of the soil has a particle size smaller than d_{85}.

In the field situation there will be an applied vertical stress acting on the geotextile and underlying soil. This vertical stress will tend to resist the mobilization of soil particles. Laboratory tests indicated that this stress can reduce the potential for piping, and the fact that it has not until now been considered represents an element of conservatism in the existing design methodology.

Based on soil grading curves and design equations (such as Eq.1), it is a relatively simple matter to select a geotextile that meets the design criteria, and a wide range of geotextiles are available, including some specifically designed to retain silt. What is more difficult to address is the potential variability of both the geotextiles and the soil. As a manufactured product, geotextiles have an inherent variability in properties and, in particular, the variability will depend on a number of factors including the method of manufacture and the level of quality control. This variability may be quite significant for some geotextiles and requires consideration. However of even greater significance is the potential variability of the soil itself; the soil samples from a few boreholes are frequently not representative of the potential range of variability that can occur in many natural deposits. For example, in recent investigations in Canada grain size distribution from a number of boreholes indicated that the material was a silty sand with a minimum d_{85} of 105 μm. However, a subsequent detailed investigation supervised by the author also identified the presence of relatively uniform silt with d_{85} of 60 μm. This indicates the need to either adopt a conservative design approach which recognizes the potential variability of both the soil and geotextiles or careful geotechnical field inspection of the insitu soils (prior to geotextile placement) to ensure that they confirm with design assumptions, combined with quality control checks on the geotextile; which of these approaches is adopted will often be based on consideration of the relative economics of the two approaches.

While the foregoing considerations are important, they are of little value if the installation of the geotextile is such that either the geotextile is torn/punctured in a manner which would permit the migration of fines and piping adjacent to the puncture/tear. Thus the selected geotextiles must not only have adequate filtration characteristics, it must also be sufficiently robust to withstand the installation; this may be particularly critical in the case when the geotextile will be in contact with coarse stone, as in Figures 1 and 2.

An area requiring special attention is the overlap/seam between adjacent sheets of geotextile. Where geotextiles are to be overlapped, an overlap of around 0.3-0.45 m is often specified. However even 0.45 m of overlap may not be adequate in critical situations such as that being discussed here. In two recent investigations in which the author was involved there was substantial evidence that there was a gap between strips of geotextile after placement even though the overlap was specified as 0.3-0.45 m. In at least one case this specification was met prior to placing the overlying material. Notwithstanding this, careful excavation of the geotextile showed an opening between geotextile sheets of up to 0.45 m! For the situation being examined here, failure is expected if a gap forms between geotextile sheets. If seams are sewn then the strength of the seam must be adequate to withstand the lateral forces that develop during construction. Thus although the design and selection of a suitable geotextile is important in these applications, careful attention to construction details is critical.

3. Geotextiles in leachate collection systems for landfills

The primary leachate collection system is a critical part of the design of the barrier system for landfills. It controls the leachate head acting on the barrier (and hence it is a key factor controlling the advective flow through the barrier). It also removes contaminants from the landfill, thereby reducing the contaminating lifespan (see Rowe, 1991a).

Figures 3-5 show schematics of three of the many possible configurations of leachate collection system details.

The differences between the three designs primarily relate to the use of geotextiles and different combinations of type of stone. There are five primary design considerations:

(a) providing adequate structural support for the pipe to minimize potential damage to the pipe;
(b) minimizing the leachate head on the landfill liner;

(c) providing for the rapid transmission of the leachate to the collection pipes;
(d) minimizing the potential for the clogging of the leachate collection system;
(e) to the maximum extent that is possible, controlling the location of the clogging which does occur, and designing to minimize the effects of clogging.

The potential modes of 'failure' for a leachate collection system include the following:

(a) breaking of the collection pipes (including long term breaking of pipe due to symbiotic effects of damage due to construction, cleaning, creep, weight of overlying waste, and differential settlements);
(b) chemical and biological clogging of the pipe and/or the pipe perforations;
(c) physical blocking of the drainage layer due to intrusion of waste from above and/or clay from below;
(d) biological, chemical and particulate clogging of the geotextile filter/separator;
(e) biological, chemical and particulate clogging of the stone used as the drainage layer.

The intent of placing a geotextile between the waste and the drainage layer is to have the geotextile function as a filter. To be judged as a good filter the geotextile has to provide for: adequate flow, upstream particle retention, and protection against clogging. It is relatively simple to select a geotextile capable of meeting the first requirement because this is controlled by parameters related to the physical properties of the geotextile. These properties are readily available from the product manufacturer. Design for filtration is more difficult because the particle size in the waste (to be retained) is uncontrolled and to a large extent unknown. Thus the objective of the filter design should be to filter out those particles which would be likely to 'settle out' in the leachate collection system. This is governed by the design of the drainage blanket and pipe system (which the engineer controls) and the characteristics of the waste (over which the engineer has little control). The ability of the geotextile to meet the third requirement, the provision of adequate protection against clogging, is difficult to assess for a geotextile which will be exposed to a landfill leachate. The high micro-organism content of leachate is known to cause clogging (as discussed below). Because of the many unknowns related to the clogging mechanism, Koerner (1990) has suggesting performing laboratory field simulation tests using site-specific materials to assess a geotextile's resistance to clogging. Recognizing that, while these tests may screen clearly unacceptable material, they do not address long term potential clogging. Thus it is desirable to recognize that clogging may occur and to design the collection system such that the potential consequences and the potential impact of geotextile clogging may be minimized.

It is well established that sand 'drainage' blankets readily clog due to biological/chemical clogging (e.g. the sand blanket at the Keele Valley Landfill in Canada), there is also some evidence to suggest that the pea gravel drainage layer at the Brock West landfill in Canada may have clogged and resulted in the development of a significant leachate mound. The likelihood of clogging can be reduced by:

(a) increasing the flow rate once leachate enters the drainage layer (this reduces residency time and reduces the degree of settling that will occur); thus the steeper the slope and/or higher the hydraulic conductivity of the drainage layer the better.
(b) the potential for biological induced clogging is related to the surface area available for biofilm growth. The potential for biological clogging is reduced by using as large diameter stone as possible.
(c) the potential for biological and chemical clogging is related to the void size. The larger the void size, the more difficult it is for clogging to occur. Generally, void size will be larger for a relatively uniform large diameter stone and will be minimal for a well graded sand or sand and stone mixture.

Brune et al. (1991) have shown that precipitation of sulphides and carbonates onto the surface of methorganic and sulphate-reducing bacteria results in clogging of stone drainage systems, waste and geotextiles. These experiments involved relatively large flows which correspond to hundreds of years of vertical infiltration through the drainage layer, but for a spacing of lateral drains of, say, 50 m and a 0.3 m stone layer it only corresponds to a few years to the order of a decade of lateral flow in the leachate collection system. It is not possible to evaluate the effect of the flow rate from this study. Below are summarized some of the conclusions that can be drawn from this study.

Clogging is related to the supply of iron and calcium in the leachate. High strength leachate may cause severe clogging in a short period of time (less than a year) in laboratory column experiments.

At the end of laboratory experiments which had been conducted for up to 15 months:

(a) drainage material with pore sizes in the range 1-32 mm and 2-4 mm were largely clogged and 'suffered almost complete loss of permeability' (certainly less than 10^{-5}cm/s)

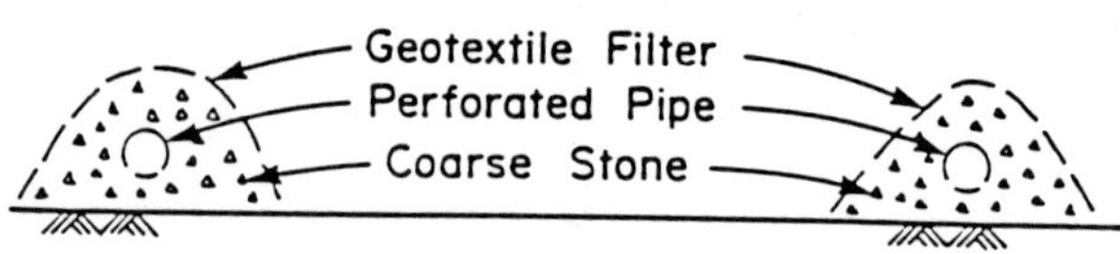

Fig. 3. Example of an inappropriate use of geotextiles for the design of a leachate collection system

(b) gravel with grainsize 8-16 mm showed a significant reduction in pore volume due to biofilm growth
(c) gravel with grainsize 16-32 mm was not clogged although biofilm growth had started. It is not clear as to how much clogging may have occurred if the tests were run for a longer time.
(d) geotextile filter layers (and the organic matter above the geotextile) experience significant clogging in high strength leachate. This is well known from other studies (e.g. Cancelli & Cazzuffi 1987); however for nonwoven needle-punched geotextiles the maximum reduction in permeability reported in the literature is to about 10^{-7}cm/s.
(e) the geotextile provided protection to the underlying stone (16-32 mm) reducing biofilm growth, compared with stone which was not protected by geotextile.

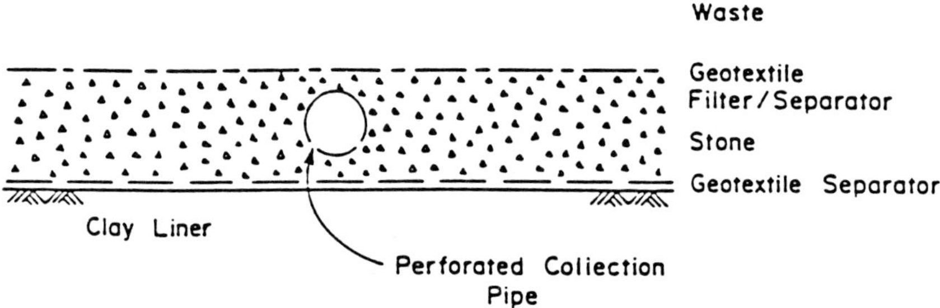

Fig. 4. Use of geotextiles as a filter/separator directly between the waste and a stone drainage layer

The three schematic designs shown in Figures 3 to 5 relate to various aspects of the issues discussed above.

Figure 3 shows a system of 'drainage' pipes surrounded by stone (or this could simple be a 'French drain' without the pipe) with a geotextile filter between the stone and the waste. This is a poor design since the flow of leachate into the drain is over a restricted area (increasing the chance of clogging in a situation where clogging of the geotextile and/or stone will greatly reduce the inflow of the drainage pipe and hence increase leachate mounding. With this design, any reduction in the efficiency of the drainage system will lead to a direct increase in the leachate head acting on the base of the landfill (e.g. the liner) and hence increase the contaminant migration through the liner. Geotextiles should not be used in this application and isolated drains such as this are not recommended.

Figure 4 shows a geotextile that separates the waste from stone and serves to protect the stone. Clogging of the geotextile is to be expected, however, and this may result in some perching of leachate in the waste. The head does not act directly on the landfill base. The effect of clogging of the geotextile can be illustrated by some examples.

(a) If a 3 mm thick geotextile clogs to a $k \approx 10^{-7}$ cm/s, a perched mound of only 16 mm above the geotextile would be required to transmit an average annual infiltration of 0.2 m per year.

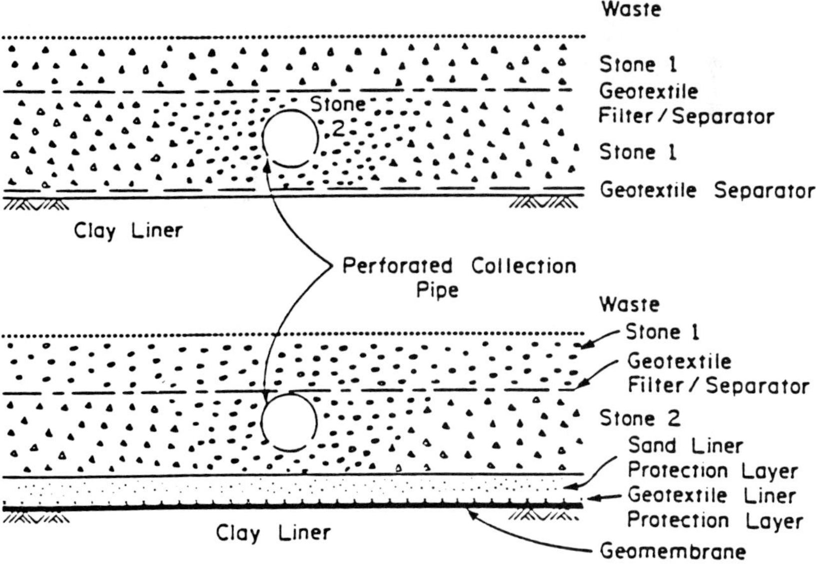

Fig. 5. Examples of multi-layered leachate collection system design including 'sacrificial' geotextile filter layer

(b) If a 150 mm thick zone above the geotextile clogs (with the geotextile) to about 10^{-7}cm/s, a perched mound of about 0.8 m would be required above the clogged zone to transmit 0.2 m per year infiltration to the underlying stone.
(c) If a 150 mm thick zone clogged to 10^{-8}cm/s, a perched mound of about 9.4 m would be required to transmit 0.2 m per year infiltration to the underlying stone.

This design reduces the risk of clogging of the stone (and consequently buildup of head on the base of the landfill but could potentially create perched conditions within the landfill. The requirement that the pipe be structurally stable for the pipe may place limits on the size and grading of the stone.

Figure 5 shows a multilayer stone system separated by a geotextile filter. In this design, it is recognized that clogging of the geotextile is likely to occur and that some perching of leachate will result. The upper stone layer (which should consist of stone in the 16 mm - 32 mm diameter range) provides some drainage above the geotextile and would be hydraulically connected to the leachate collection manhole. Thus in this design, the upper stone/geotextile is intended to provide some drainage and protection of lower stone and to reduce the problem of clogging of a thick zone above the geotextile. Large open stone (e.g. 50 mm clear stone) is used away from the pipe in the (lower) primary drainage layer. Smaller stone (required for pipe stability) is used near the pipe, however the pipe should be designed to allow the use of at least 16 mm diameter (or larger) stone adjacent to the pipe in order to minimize clogging of this stone.

Thus although geotextiles have been criticized because of the potential for clogging, it must be recognised that this potential exists for all the materials used in the drainage system. The objective of the design of a leachate collection system should be to develop a system which will minimize the buildup of leachate head on the base of the liner for as long as practical. Geotextiles can perform a very useful role in the design of these collection systems, provided the potential for clogging is recognized and accounted for in the design of the system.

4. Conclusion

This paper has highlighted the important applications for geotextiles as filters in situations where there is significant potential for failure of the system in the event that the design or construction is inadequate. The primary objective of the paper is to provide a basis for initiating discussion of a number of issues that arise from these examples. These include

(a) the significance of the potential variability of the filtration characteristics of the geotextile and the grainsize distribution of the soil it is intended to protect
(b) the importance of construction specifications and quality control; in particular, the need to prevent opening of seams and the consequent gap between geotextile sheets
(c) the potential for clogging of geotextiles (especially in landfill applications) and the need to design systems which take account of this potential clogging.

5. Acknowledgements

This paper arose from a general programme of research in to geosynthetics in both conventional geotechnique and geoenvironmental applications being funded by the Natural Science and Engineering Research Council of Canada under grant #A1007.

6. References

BRUNE M., RAMKE H.G., COLLINS H.J. and HANERT H.H. (1991). Encrustation processes in drainage systems of sanitary landfills. *Proceedings Sardina '91: Third International Landfill Symposium*, S. Margherita di Pula, Cagliari, Italy. October 14-18, pp. 999-1035.
CANCELLI A. and CAZZUFFI D. (1987). Permittivity of geotextiles in presence of water and pollutant fluids. *Proceedings Geosynthetics '87*, New Orleans, pp. 471-481.
FAURE Y., GOURCE J.P., BROCHEIR P., and ROLLIN A.L. (1986). Soil - geotextile interaction in filter systems. *Proceedings 3rd International Conference on Geotextiles*, Vienna.
GIROUD J.P. (1982). Filter criteria for geotextiles. *Proceedings 2nd International Conference on Geotextiles*, Las Vegas, pp. 103-108.
KOERNER R.M. (1990a). Preservation of the environment via geosynthetic containment systems. *4th International Conference on Geotextiles and Geomembranes*, The Hague.
ROWE R. K. (1991a). Contaminant impact assessment and the contaminating lifespan of landfills. *Canadian Journal of Civil Engineering*, volume 18, pp. 244-253.
ROWE R. K. (1991b). Leachate detection or hydraulic control: two design options. *Proceedings 3rd International Symposium on Sanitary Landfills*, Sardinia, pp. 997-987.
ROWE R. K. (1991c). Some considerations in the design of barrier systems.

Proceedings 1st Canadian Conference on Environmental Geotechnics, Montreal, pp. 157-164.
ROWE R.K. (1992). Integration of hydrogeology and engineering in the design of waste management sites. *Proceedings of the Conference of International Association of Hydrogeologists,* Hamilton.

Geotextiles in the design of filtration and drainage for highways

D. M. FARRAR, Transport Research Laboratory, Crowthorne, Berks, UK

This paper considers geotextiles in filtration and drainage design for highway applications. These applications are described, and illustrated with case histories. Design methods are based mainly on laboratory index tests in the BS 6906 series. In addition to describing design procedures, consideration is given to other factors which influence the drainage and filtration performance, such as site activities, access and maintenance. The paper reviews the limitations of current methods of property testing and design, and briefly discusses the future use of classification and attestation systems.

1. Introduction

Filtration and drainage play an important role in many geotechnical applications, and geotextiles are being increasingly employed either as filters in conjunction with granular materials or pipes, or as geocomposites (fin drains) with both filtration and drainage roles. Geotextiles have now been used for some twenty years; an OECD review summarises the position in 1990 (Murray 1991).

The paper considers design for pavements, earthworks and earth retaining structures for highways. Although a major application by itself, many of the considerations are also relevant in such other fields as agriculture, railways and building construction. There are, however, important applications where additional considerations apply which are not addressed in this paper. These include erosion control (reversing and agitated flow: Hemphill and Bramley (1989)), water-retaining structures and dams (long life and high factors of safety), and waste retention (very high standards of installation and guaranteed long-term performance).

Geotextiles in filtration and drainage. Thomas Telford, London, 1993

2. Highway applications, and experience

The main uses of geotextiles in filter and drainage design for highways are summarised below, and some studies which have been made of these uses are also described.

2.1. Edge of pavement drainage

It has long been recognised that excess water in the pavement structure can be deleterious, and that adequate drainage is necessary (Cedergren 1974). Edge of pavement drainage may consist of a combined surface and sub-surface system (French drains), or separate systems (DTp 1987). An estimate must be made of the surface and sub-surface water entering the drain at different stages of the design life of the pavement. The grading of the pavement materials and adjacent soils, and of road detritus for combined systems (Samuel and Farrar 1988), must also be considered. Subsequent access for maintenance or repair is possible for French drains, but may be more difficult for some types of sub-surface drain.

Farrar and Samuel (1989) describe observations of a geotextile filter in a French drain (Terram 700 wrapped round a coarse filter material). Tests on recovered samples of geotextile showed no sign of deterioration after 5 years in service, although there may have been some damage during installation. The study showed that the geotextile was effective in retaining roadside detritus, but that remedial measures were necessary to remove the accumulation of detritus which was impeding flow.

2.2. Slope drainage

Drainage is only one of several possible remedial measures that may be considered to ensure stability of slopes, and cost is therefore an important factor in selecting the best option (Johnson 1985). Applications range from relatively shallow slope drains that play a secondary role in stabilising the slope or are designed on superficial investigation, to major works intended to stabilise deep slips and based on a full site investigation.

Farrar (1990) has described observations on slope drains incorporating a geotextile filter (Typar 3407) wrapped round granular material installed in an old cutting in London Clay. After seven years in service, there had been no deterioration in performance, and tests on recovered samples of the geotextile also showed no signs of deterioration. However, a wet zone and some alterations in grading were observed in the clay adjacent to the geotextile.

2.3. Drainage blanket

The purpose of a drainage blanket is to dissipate excess pore pressure within or beneath an embankment or other structure. Satisfactory performance is critical to stability, although the blanket is usually only required to function for a year or two. The drainage requirements can usually be reliably determined after a site investigation. Finlayson *et al.* (1984) describe a major failure involving a clay embankment over a drainage blanket which demonstrated the importance of careful consideration of the design requirements in this application.

2.4. Use of fin drains

Fin drains are being increasingly used in highway applications as a means of conserving imported granular drainage materials and of simplifying construction procedures. It is necessary, however, to ensure that they can be properly designed and installed.

Samuel *et al.* (1987) have studied the long-term performance of lengths of fin drains (Filtram and Trammel) laid in 1979, in a shallow cutting in heavy clay at the side of a minor road. In spite of some deficiencies in installation, the fin drains were found to be performing as effectively in 1985 as in 1979. There was no evidence of deterioration in recovered samples of geotextile apart from calcareous deposits at one level.

Corbet (1990) carried out a full scale trial of five types of fin drain (Enkadrain, Filtram, Hidrain, Hydraway, Trammell) with two objectives; firstly to establish the problems likely to arise on-site in installing fin drains for edge of pavement drainage, and secondly to compare their in situ performance with that predicted from the index tests presently available for geotextiles. The relevant conclusions which were drawn from this trial were as follows.

(a) The fin drains were installed against a gap graded sand containing silt and clay fractions and selected as a difficult soil. Over two days, the flow through the soil/geotextile interface was found to be about one thousandth of that measured in laboratory index tests on the geotextile filter alone.

(b) Four of the five geotextile filters functioned effectively in this soil, and one failed. A number of published design criteria successfully predicted this result. The study also indicated that criteria for such applications as river protection, which involve reverse flows, are probably too severe for uni-directional flow situations.

(c) In the trial, in situ flow in the plane of the fin drain, away from the inlet and outlet, was found to be of the same order as that measured in laboratory index tests at a comparable hydraulic gradient. However, the flow regime in fin drains is not well understood.

2.5. Retaining structures

Drainage is an essential consideration in the design of earth retaining structures (DTp 1987b). Filter drains are in principle required to function with limited or no access for periods of up to 120 years, and may be subjected to large normal and shear stresses. The requirements for fill materials are well defined. The properties of any natural ground should be known from site investigation. There is, however, a lack of case histories on the long term performance of drainage to highway structures. In a review of this subject, Bird (1992) concluded that there was no evidence of failures due to drainage in structures built to recent design standards. It appears, however, that designers are unsure of what assumptions to make on hydraulic pressures. The absence of failures therefore may merely reflect conservative assumptions in design.

Geotextiles and fin drains have been used on a number of structures and appear to be functioning successfully, although full-scale monitoring of a retaining structure would be desirable. The cost of drainage is a very small part of the overall cost of such structures, typically 1-2%, and clearly the emphasis is on the long term performance of the entire structure, and not just on initial costs of the drainage component.

3. Filter and drainage design

3.1. Objective of design

The objective is to ensure stability throughout the design life by a reduction in pore pressure or depth of water table. The steps required in any design are summarised below.

(a) Define the stability requirement and design life. For example, different configurations of slope drain would be required for the prevention of deep circular slips and of shallow planar slips.
(b) Assess the reduction in water table or pore pressures required, and the drainage configuration required to achieve this. Methods employed include design charts, flow nets, and finite difference or finite element methods (e.g. Hutchinson 1977, Cedergren 1989).
(c) Based on the hydraulic gradient and permeability of the soil,and an

estimate of surface water entering the system, the quantity of water to be drained must then be determined. Soil permeability may be determined by measurement, but is usually estimated from published figures or the grading of the soil (Kenney *et al.* 1984). It will often only be possible to estimate permeability to an order of magnitude, and a conservative estimate may be desirable.

(d) Finally, filtration must be considered, in order to ensure that the drainage function is maintained throughout the design life. This involves consideration of all the soils in contact with the geotextile.

If the geotextile is in contact with several different soils or fills, it will be necessary to check the filtration and flow requirements for each soil. If the requirements are incompatible, it may be necessary to consider employing more than one geotextile.

3.2. *Test procedures*

In the UK, design methods are based at present on the BS 6906 series of index tests, with a few additional index tests from other sources. These tests are likely to be superseded by a more comprehensive series of CEN index tests within a few years but, where there is presently a UK equivalent, the CEN test will probably not differ greatly.

3.3. *Durability*

Geotextiles will be in contact with natural soil and fill, specified industrial waste products, and cement- and lime- stabilised materials. Ground water may contain such contaminants as road salt and diesel oil. There will also be a short-term exposure to UV radiation. In the present absence of suitable tests, durability is specified in such general terms as 'The Contractor shall provide evidence that the geotextile ...will be sufficiently durable ...'. Specialist advice will be needed if a geotextile is to be employed for a waste containment purpose.

3.4. *Filtration*

The geotextile must retain soil particles (i.e. prevent them clogging the pores), and permit the passage of water (i.e. prevent blinding or blocking of the geotextile surface). The design of a geotextile filter is usually based on three parameters.

(a) The uniformity coefficient of the soil (U= D60/D10)
(b) The mean particle size of the soil (D50)
(c) The apparent pore size (O90) of the geotextile

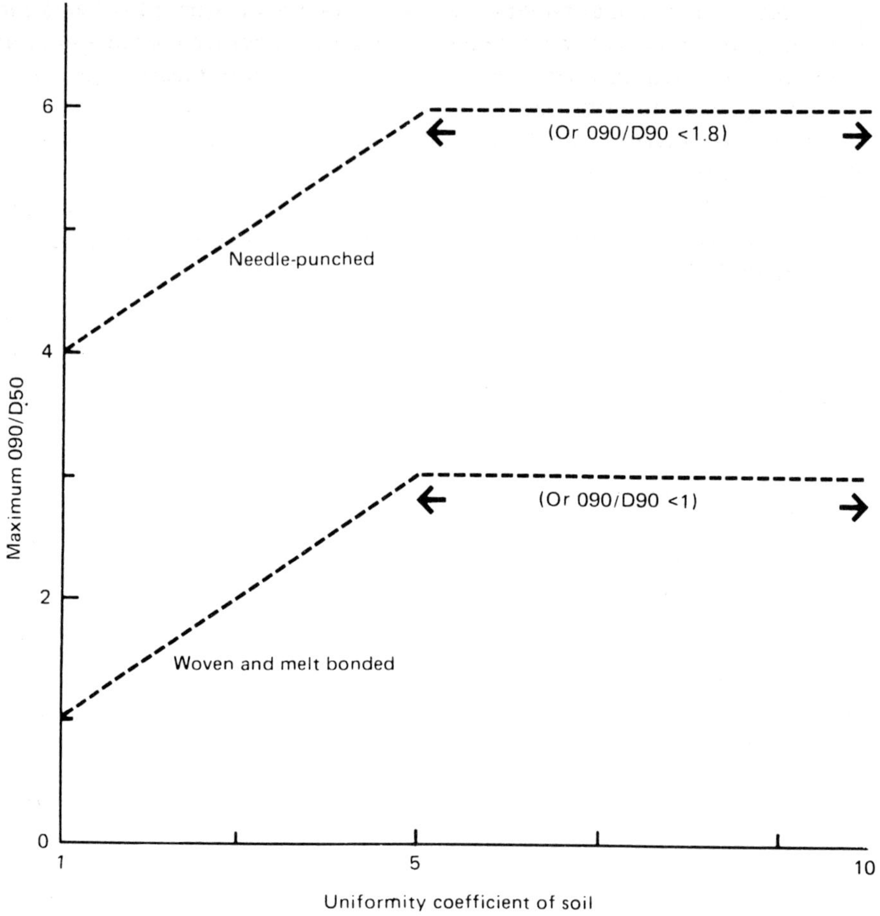

Fig. 1. Filtration criteria for geotextile criteria

(where Dn is a soil particle size for which n% by weight is smaller than that size, and O90 is the apparent pore size of a geotextile, i.e. that for which 90% of the particles of that size used in the sieving test are retained on the geotextile).

O90 is presently determined by dry sieving (BS 6906: Part 2); this is not satisfactory for some geotextiles, and until CEN tests are available it will be necessary to accept results from other test procedures which involve wet sieving.

Numerous criteria have been proposed for the relation between the above parameters (e.g. VanZanten 1986, John 1987). The relations due to McGown *et al.* (1982) and shown in Fig. 1 are suggested in the Note for Guidance to the new DTp Specifications (DTp 1992b), to obtain maximum design values of O90 for highway applications.

In fine grained soils, these criteria may lead to excessively low flow rates through the geotextile. In some coarse soils, the criteria may lead to problems with the internal stability of the soil. Limits should therefore be placed on design values; 0.1 and 1mm respectively in DTp Notes for Guidance. If the soil is gap-graded, the coarser fraction is neglected in the calculation. A soil may be regarded as gap-graded and potentially unstable if the finest 30% does not meet the criterion (Kenney and Lau 1985)

$$W(4D) > 2.3W(D) \tag{1}$$

(where W(D) and W(4D) are the percentage weights of particles smaller than given diameters D and 4D respectively).

Designers should be aware that a thin layer of weak soil may form at the geotextile/soil interface. In any stability calculation, it would seem prudent to assume the critical angle of friction (ϕ_{cv}) for this layer.

3.5. *Flow normal to plane of geotextile*

BS test BS 6906: Part 3 determines the flow rate per unit area (F) normal to the plane of a geotextile under a standard head (h_g) of 0.1m. If the flow is assumed to be laminar, the permeability (k_g) may be obtained from the relation

$$k_g = Ft_g/h_g \tag{2}$$

(where t_g is the thickness; if required, this may be determined by ISO test ISO: 9863 (1990)).

The BS test also measures a breakthrough head, although this is not usually considered in highway applications. Either the measured permeability or flow rate may be used in design.

Permeability. This property is expressed by the inequality

$$k_g > nk_s \tag{3}$$

where k_s is the permeability of the adjacent soil or fill; and n is a factor allowing for the difference between the behaviour of the geotextile in isolation and of the geotextile/soil combination (it should strictly be site-specific but is taken to apply generally with sufficient accuracy). n may be determined from relations due to Heerten (1984) and reproduced in John (1987) and

Table 1. Effect of softer surface on transmissivity

Surface	Reduction in transmissivity	
	Mean	Range
Hard rubber	13%	0-24%
Soft rubber	26%	6-46%

(Note; 'Hard' rubber is defined as having a compression of 7% at 100kPa, BS 903:1990. 'Soft' rubber is defined as having a compression of 39% at 100kPa)

Hemphill and Bramley (1989). It will have values differing for wovens and non-wovens and increasing from about 10 to 1000 with decreasing particle size of the soil.

Flow. The flow through the geotextile and adjacent soil are compared using the inequality (CFGG 1989)

$$k_g h_g / t_g > n' k_s i_s \quad (4)$$

where i_s is the hydraulic gradient in the adjacent soil.

The factor n′ is the product of several factors (type of structure, compression under load, hydraulic gradient in soil, head loss in geotextile). In highway applications, it is reasonable to take h_g as about 0.1m and i_s as less than 10. Equation (4) then becomes

$$k_g / t_g > n' k_s \quad (5)$$

for units of metres and seconds.

There is clearly a wide range of design factors for flow; in the absence of better information, a value of n of 100 and of n′ between 2×10^4 and 10^5 might be appropriate. Values of n′ of 10^4 are suggested by the CFGG in high-risk and 10^3 in lower risk situations for most soils (equivalent to a value of n of between about 1 and 50). In well defined applications, it may be possible merely to specify a minimum flow through the geotextile; e.g. the DTp suggests a minimum F of 30 litres/s/m^2 for edge of pavement drainage (DTp 1992b).

3.6. *Flow in plane of geotextile*

BS test 6906: Part 7 is used to determine the relation between hydraulic gradient (h) and flow per unit width in the plane of the geotextile (q_{pg}). If flow is laminar, the transmissivity (k_{pg}) may then be obtained from the relation

$$k_{pg} = q_{pg}/h \tag{6}$$

The BS test recognises that flow is not laminar, and that transmissivity should be quoted for a hydraulic gradient appropriate to the application. The test is a short-term one, under immediately applied load. Two factors must be taken into account in estimating the design flow in plane.

First, the fin drain will be in contact with soil, and this may deform the geotextile filter stretched over the plastic core and reduce the cross-sectional area available for flow. The BS tests gives the option of using softer rubber or soil surfaces in contact with the specimen, as an alternative to hard surfaces, to simulate this effect. Tests on three fin drains (Filtram, Hydraway, Hitek) gave the results shown in Table 1 for softer surfaces under compressive stresses of up to 100kPa.

Second, the fin drain will be subject to compressive creep under normal or shear loading due to transient compaction stresses during installation and to subsequent long-term overburden stresses. It is therefore necessary to determine the long term thickness under compressive creep, and then simulate this condition in the short-term transmissivity test (Murray and McGown 1992). Two examples of compression curves under a normal load of 50kPa and shear load of 5kPa are shown in Fig. 2. Fin drain A has a mesh core; it

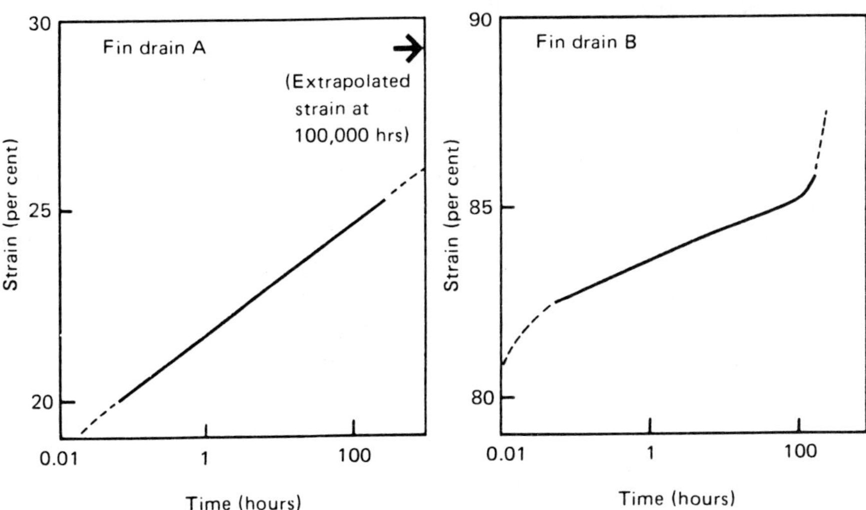

Fig. 2. Examples of compressive creep for fin drains

has a consistent creep, but is likely to retain some 70% of its original thickness throughout its design life. In contrast, fin drain B has a weak random core; after both initial and long term compression it only retains a fraction of its original thickness. There is presently no standard test for compressive creep, and the results shown in Fig. 2 were obtained by the Transport Research Laboratory using a modified oedometer with stresses applied at an angle to the plane of the fin drain.

Compaction stresses will depend on the method of installation of the geotextile, but for bulk earthworks typically correspond to a vertical stress of 250kPa and a lateral stress of 50kPa (Bolton 1991). Long term vertical stress will probably be the overburden stress, and long term lateral stress will lie between the K_a condition and the lateral overburden stress enhanced by compaction, depending on the amount of yielding that occurs. Shear stresses on the geotextile will be about half of these values.

Flow will be downward to the lower edge of the installed fin drain. In fin drains which do not incorporate a collector pipe, there will also be a near horizontal flow along the length. A design factor of about 2 is suggested if account is taken of the points discussed above, otherwise a factor of at least 10 would be appropriate.

3.7. Other considerations

The engineer should also be aware of the following factors which may be relevant to design, although strictly these should be covered in the specification

Installation damage. Geotextiles must resist installation stresses, and accommodate any irregularities in the surface on which they are laid. They must also withstand impact and abrasion caused by any fill dumped and compacted over them. DTp Specification clauses (DTp 1992) give design values, derived from histories of successful usage of geotextile filters, of

Maximum load per unit width (BS 6906: Part 1) > 5kN/m
Strain at breaking load (BS 6906: Part 1) > 10%
Puncture resistance (BS 6906: Part4) > 1.2kN
Tear resistance (ASTM D4533-85) > 200N

It may be noted that there is doubt about the general applicability of the tear test. Moreover there is no generally accepted test for abrasion resistance.

Ease of installation. It must be feasible to install the filter or drain under site conditions. Consideration should be given to ease of installation, the possibility of misuse or damage during later stages of construction, and methods of checking that it is functioning correctly at the end of construction. The specification should ensure that joints between lengths and to outfalls are appropriately designed.

Maintenance and repair. The possibility of subsequent access for inspection, maintenance and repair is an important consideration which should be addressed at the design stage. Particular requirements could include; the provision of weep holes or other means of minimising the effect of any subsequent blockage, access chambers, provision for rodding or inspection of pipes, and periodic removal of surface water detritus. Simple but nevertheless informed inspection procedures will minimise the possibility of expensive failures, e.g. by identifying a blocked drain before the slope or pavement with which it is associated fails. If subsequent access is possible, the design life of the drainage may be less than that of the structure with which it is associated.

4. Problems associated with design and testing

The main problems remaining in testing and design of geotextiles for filtration and drainage are listed below. All are now being addressed, mainly under the impetus of European harmonisation.

4.1. *Durability*

There is presently no readily available range of durability tests, although a British Standards review (1992) will be helpful in providing guidance. This subject will become increasingly important with the development of European harmonisation of products. Damage tests are being developed for reinforcement applications, and tests for other applications would be desirable.

4.2. *Pore size*

The introduction by CEN of improved methods for pore size determination will improve matters both from a design and contractual viewpoint. A better understanding of pore structure (Gourc and Fauré 1992), possibly with novel test methods such as mercury intrusion (Praparahan *et al.* 1989), may lead to improved design methods. It may be desirable to measure pore size under stress.

4.3. *Normal flow*

Because of the huge disparity between index test and actual flows, there seems little point in attempting to refine this test, except possibly by the introduction of applied stress.

4.4. Behaviour of soil/geotextile interface

There is a great need for performance tests which will accurately predict the long term flow and filtration of a geotextile in contact with a soil. A number of tests have been proposed, such as the gradient ratio test, but none appear wholly satisfactory (Koerner and Ko 1982, Williams and Abouzakhm 1989, Fourie and Bentley 1990). Suitable performance tests could either be used directly, or used as an aid to 'calibrate' design rules based on index tests (Lawson 1986).

4.5. In-plane flow

An index test such as British Standard BS 6906: Part 7 may prove to be an adequate basis for design, although more needs to be known about the simulation of the effect of compressive creep and the hydraulics of in-plane flow.

4.6. Clogging

For applications where the filter must function over many decades without attention, more must be known about the conditions leading to chemical or biological clogging, i.e. the combinations of air, chemicals, temperature and flow most likely to produce these difficulties

4.7. Classification

Geotextiles could be classified by

(a) Product: the product alone is classified without reference to the application; e.g. strength classes for pipes.
(b) Application: both the product and its specific application are defined; e.g. pipes for drainage in DTp applications (DTp 1987).

UK producers and users need to consider if such systems would be helpful to them.

4.8. Attestation

A highway authority will probably require an attestation system with some degree of independent checking for geotextiles and other products, to accompany public procurement procedures. UK producers and users must consider their approach to this question.

5. Discussion

The design methods described in this paper, although largely empirical, are adequate at least for situations where the consequences of failure are not too severe and the input for very accurate design is often unavailable. These methods are also used for situations involving drainage of major earthworks and structures, where the geotextile may be required to perform under significant stresses over long periods with limited or no access. Even for simpler applications, the engineer should always be prepared to modify design methods in the light of experience of particular combinations of soil and geotextile.

The next few years are likely to provide challenges and opportunities in increasing the range of applications for geotextiles, and in the integration and harmonisation of European Standards. It is hoped that progress in dealing with the problems listed in the paper will enable designers to meet these changes. Such changes are likely to increase the pressure for better design methods and more reliable test procedures, particularly for performance properties needed in design.

6. Acknowledgements

The work described in this paper forms part of a Department of Transport funded research programme conducted by the Transport Research Laboratory, and the paper is published by permission of HE and BE Divisions and the Chief Executive of TRL.

7. References

AMERICAN SOCIETY FOR TESTING AND MATERIAL (1985). ASTM D4533-85. *Standard test method for trapezoid tearing strength of geotextiles.*

BIRD R.W. Review of drainage to earth retaining structures to highways (in preparation).

BOLTON M.D. (1991). *Geotechnical stress analysis for bridge abutment design.* Dept of Transport TRRL Contractor Report 270 (Transport and Road Research Laboratory, Crowthorne).

BRITISH STANDARDS INSTITUTION (1990). BS 903: Part A4: 1990. *Determination of compressive stress-strain properties.* British Standards Institution, London.

BRITISH STANDARDS INSTITUTION (1987). BS 6906: Part 1: 1987. *Determi-*

nation of the tensile properties using a wide width strip. British Standards Institution, London.

BRITISH STANDARDS INSTITUTION (1989). BS 6906: Part 2: 1989. *Determination of the apparent pore size distribution by dry sieving* (British Standards Institution).

BRITISH STANDARDS INSTITUTION (1989). BS 6906: Part 3: 1989. *Determination of water flow normal to the plane of the geotextile under a constant head* (British Standards Institution).

BRITISH STANDARDS INSTITUTION (1989). BS 6906: Part 4: 1989. *Determination of the puncture resistance (CBR puncture test)* (British Standards Institution).

BRITISH STANDARDS INSTITUTION (1990). BS 6906: Part 7: 1990. *Determination of in-plane waterflow* (British Standards Institution).

CEDERGREN H.R. (1974). *Drainage of highway and airfield pavements* (Wiley).

CEDERGREN H.R. (1989). *Seepage, drainage and flow nets* (Wiley).

COMITÉ FRANÇAIS DES GÉOTEXTILES ET GÉOMEMBRANES (1989). *Recommendations for the use of geotextiles in filtration and drainage systems* (CFGG)

CORBET S.P. (1990). *Comparative trials of fin drains.* Dept of Transport TRRL Contractor Report 221 (Transport and Road research Laboratory, Crowthorne).

DEPARTMENT OF TRANSPORT (1987). *Highway construction details.* (HMSO 1987).

DEPARTMENT OF TRANSPORT (1987b). *Backfilled retaining walls and bridge abutments.* Departmental Standard BD 30/87 (Dept of Transport).

DEPARTMENT OF TRANSPORT (1992). *Specification for highway works,* 2nd Edn (HMSO, in preparation).

DEPARTMENT OF TRANSPORT (1992). *Notes for Guidance on the Specification for Highway Works,* 2nd Edn (HMSO, in preparation).

FARRAR D.M. and SAMUEL H.R. (1989). *Field study of filter drains at Swanley, Kent.* Dept of Transport TRRL Research Report 179 (Transport and Road Research Laboratory, Crowthorne).

FARRAR D.M. (1990). *Observations of slope drainage at Romford, Essex.* Dept of Transport TRRL Research Report 302 (Transport and Road Research Laboratory, Crowthorne).

FINLAYSON D.M., GREENWOOD J.R., COOPER C.E. and SIMONS N.E. (1984). Lessons to be learnt from an embankment failure. *Proc. Instn. of Civil Engrs.,* **76,** Part 1, 207-220.

FOURIE A.B. and BENTLEY N.G. (1990). A laboratory study of factors affecting performance of geotextile-wrapped drainage pipes. *Geotextiles and Geomembranes,* **10,**1-20.

GOURC J.P. and FAURÉ Y.H. (1992). Soil particles, water ... and fibres - A fruitful interaction now controlled. *Proc. 4th Internat. Conf. on Geotextiles,*

Geomembranes and Related Products, The Hague, The Netherlands, pp949-971 (Balkema)

HOFFMAN G.R. and TURGEON R. (1983). *Long-term in situ properties of geotextiles.* Transportation Research Record 916, 89-94.

HEERTEN G. and WITTMAN L. (1984). Filtration properties of geotextiles and mineral filters demonstrated by the example of bank protection. *23rd Internat. Man-made Fibres Congress*, Austria.

HUTCHINSON J.N. (1977). Assessment of the effectiveness of corrective measures in relation to geological conditions and type of slope movement. *Bull.Internat. Assocn. of Engng. Geology*, **16**, 131-155.

INTERNATIONAL STANDARDS ORGANISATION (1990). ISO 9863:1990. *Geotextiles - determination of thickness at specified pressures* (ISO).

JOHN N.W.M. (1987). *Geotextiles* (Blackie).

JOHNSON P.E. (1985). *Maintenance and repair of highway embankments:studies of seven methods of treatment.* Dept of Transport TRRL Research Report 30 (Transport and Road Research Laboratory, Crowthorne).

KENNEY T.C., LAU D. and OFOEGBU G.I. (1984). Permeability of compacted granular materials. *Canad. Geotech. J.*, **21**, 726-9.

KENNEY T.C. and LAU D. (1985). Internal stability of granular filters, *Canad. Geotech. J.*, **22**, 215-225.

KOERNER R.M. and KO F.K. (1982). Laboratory studies on long term drainage capability of geotextiles. *Proc. 2nd Internat. Conf. on Geotextiles*, Las Vegas. Vol 1, pp 91-95.

LAWSON C. (1986). Geotextile filter criteria for tropical residual soils. *Proc. 3rd Internat. Conf. on Geotextiles*, Vienna. pp557-562.

McGOWN A., KABIR M.H. and MURRAY R.T. (1982). Compressibility and hydraulic conductivity of geotextiles. *Proc 2nd Internat Conf on Geotextiles*, Las Vegas, Vol 1, pp167-172 (Industrial Fabrics Association International)

MURRAY R.T. (Ed.) (1991). *Ground engineering applications of geotextiles in road construction and maintenance.* Report of OECD Experts Group I8 (Transport and Road Research Laboratory, Crowthorne).

MURRAY R.T. and McGOWN A. (1992). *Ground engineering applications of fin drains for highways.* Dept of Transport TRRL Application Guide 20 (Transport and Road Research Laboratory, Crowthorne).

PRAPARAHAN S., HOLTZ R.D. and LUNA J.D. (1989). *Pore size distribution of nonwoven geotextiles.* American Society of Testing and Materials, pp. 261-268.

SAMUEL H.R., FARRAR D.M., JEWELL R.A. and WALE J.H. (1987). Performance of fin drains in a road cutting. *9th European Conf. on Soil Mechanics and Foundation Engineering*, Dublin. Pp 241-244.

SAMUEL H.R. and FARRAR D.M. (1988). *Field study of filter drains on Maidstone By-Pass, Kent.* Dept of Transport TRRL Contractor Report 61 (Transport and Road Research Laboratory, Crowthorne).

VAN ZANTEN R.V. (Ed.) (1986). *Geotextiles and geomembranes in civil engineering* (Balkema).
WILLIAMS N.D. and LUETTICH S.M. (1990). Laboratory measurements of filtration characteristics. *4th Internat. Conf. on geotextiles, Geomembranes and Related Products,* The Hague. pp273-278. (Balkema)

The design and specification of geotextiles and geocomposites for filtration and drainage

S. P. CORBET, CEng, MICE, Principal Engineer, G. Maunsell & Partners, UK

Notation

O_{90} - Apparent opening size (AOS), 90% of openings are finer than value in mm
Kg - Permeability of geotextile
Ks - Permeability of soil
d_{90} - Size of soil grain size of which 90% are finer
d_{60} - Size of soil grain size of which 60% are finer
d_{10} - Size of soil grain size of which 10% are finer
D_{15} - 15% size in filter
D_{85} - 85% size in filter
D_{98} - 98% size in filter
U_c - Coefficient of Soil Uniformity d_{60}/d_{10}

1. Introduction

The use of synthetic geotextiles, fabrics manufactured from polypropylene, polyester, polyamides or polyethylene, in civil engineering works started about 20 years ago. The earliest uses of geotextiles were as separators and filter layers to help keep drains clean and to eliminate graded material filters. The filtration function has remained one of the most important uses of geotextiles, which has been extended with the development of drainage geocomposites, where the geotextile covers a polymer core.

The manufacturing process used in the production of geotextiles determines their characteristics. For particular applications which can be described as filtration or drainage some geotextiles may be better suited than others. To be able to specify the correct geotextile the engineer must have a

Geotextiles in filtration and drainage. Thomas Telford, London, 1993

knowledge or guidance as to the significance of these variations. The design which precedes specification therefore needs to be undertaken with this knowledge.

Specification of geotextiles using the time worn 'Brand X or similar approved' method based on previous experience or manufacturers recommendation is in fact restrictive. Unless the Engineer specifies which of the geotextiles properties are relevant for the situation the Contractor has no choice other than Brand X.

The quality of geotextiles should generally be good as they are produced in factory conditions often under Quality Assurance Schemes. Variations in the measured properties will always occur, any certificates showing compliance with specifications should show the variations measured. The specification should be clear as to whether the values quoted are minimum or maximum and whether average values will be accepted in compliance. Regular testing to ensure compliance with a specification must be carried out. The testing should preferably be carried out by an independent laboratory or a manufacturers laboratory regularly inspected by an independent body, e.g. British Board of Agrement (BBA).

2. Types of geotextile and geocomposite used in filtration and drainage

Geotextiles are usually described by the method of manufacture and/or the material used in their construction. The methods of manufacture usually employed are

(a) weaving: split films or tapes, monofilaments, multistrand filaments
(b) spinning: with heat bonding,with needle punching
(c) knitting: multistrand filaments

The polymers used in the manufacture can be polyethylene, polyester, polypropyleneor or polyamide (nylon).

Geocomposites used for drainage are a combination of a geotextile filter and a core through which the fluid can be transmitted to a discharge point. The cores used can be grouped as

(a) mesh nets
(b) random fibres
(c) cuspated cores (symmetrical and asymmetrical
(d) pillar cores
(e) corrugated cores (usually for band drains)

The cores are usually produced using polyethylene, polypropylene or polyamide (nylon).

3. Characteristics of the geotextiles and geocomposites

The manufacturing methods used produce geotextiles with particular characteristics both in appearance and in their engineering properties.

Table 1. Geotextile types

Method of Manufacture	Fibres/or Treatment	Characteristics in Filtration and Drainage
Weaving	Tapes	▪ Medium to high permittivity ▪ EOS can be controlled by weaving process ▪ Fibres can spread changing the EOS
	Mono filaments	▪ Relatively large EOS sizes ▪ High permittivity ▪ Fibres can spread changing the EOS
	Multifilaments	▪ Smaller EOS sizes ▪ Medium permittivity ▪ Puncture resistance better than others
Spinning	Heat bonding	▪ EOS can be variable as density of fibres varies in fabric ▪ Permittivity is variable ▪ Some fabrics have high water support head
	Needle Punching	▪ EOS can be variable ▪ EOS sizes generally smallest ▪ Can be susceptible to clogging against silts ▪ Puncture resistance is usually good, particularly for heavier fabrics
Knitting	Multi filaments	▪ EOS can be closely controlled by manufacture ▪ Puncture resistance high ▪ Permittivity can be controlled by manufacture ▪ Not common

The manufacturing process and the polymers used will determine the characteristics of the cores of geocomposites.

Table 2. Geocomposite cores

Type of Core	Polymer	Characteristics
Mesh Nets	Polypropylene or Polyethylene	▪ Low transmissivity ▪ Medium resistance to compression creep
Random Fibres	Polyamide or Polyethylene	▪ Medium to low transmissivity ▪ Best at low to medium lateral pressures ▪ Comprehensive creep can be a problem
Cuspated Cores (Symmetrical)	Polypropylene	▪ High transmissivity ▪ Can resist high lateral pressures ▪ Can collapse suddenly under load ▪ Can suffer collapse due to progressive creep under load ▪ Shear loads can affect behaviour
Cuspated Cores (Asymmetrical)	Polypropylene	▪ As symmetrical cores ▪ Can also collapse on one side reducing transmissivity on one side
Pillar Cores	Polypropylene	▪ High transmissivity ▪ Can collapse suddenly under load, particularly with shear load ▪ Can suffer collapse due to progressive creep
Corrugated Cores	Polypropylene Polyethylene	▪ Usually thin for use in band drains ▪ Can resist high lateral pressures ▪ Transmissivity low but volumes of fluid usually small

To minimise the reduction of transmissivity by the geotextile being pushed into the core under load, a well designed core is needed to provide support for the geotextile filter. Transmissivity tests, BS 6906 Part 8 [Ref. 1] modified to include soil or pseudo soils (soft rubber) can be carried out to model this effect, see Fig. 1.

4. Design of geotextile filters

The design of geotextile filters requires details of the soils which will be in contact with the geotextile, the situation in which the geotextile will be used (e.g. road drainage, railway drainage, under revetments, structural drainage, flood defences, water retaining structures) and the type(s) of geotextile which will be used or specified.

The two criteria to be considered in the design of geotextile filters are: soil retention and permeability. For use in road pavement and structural drainage

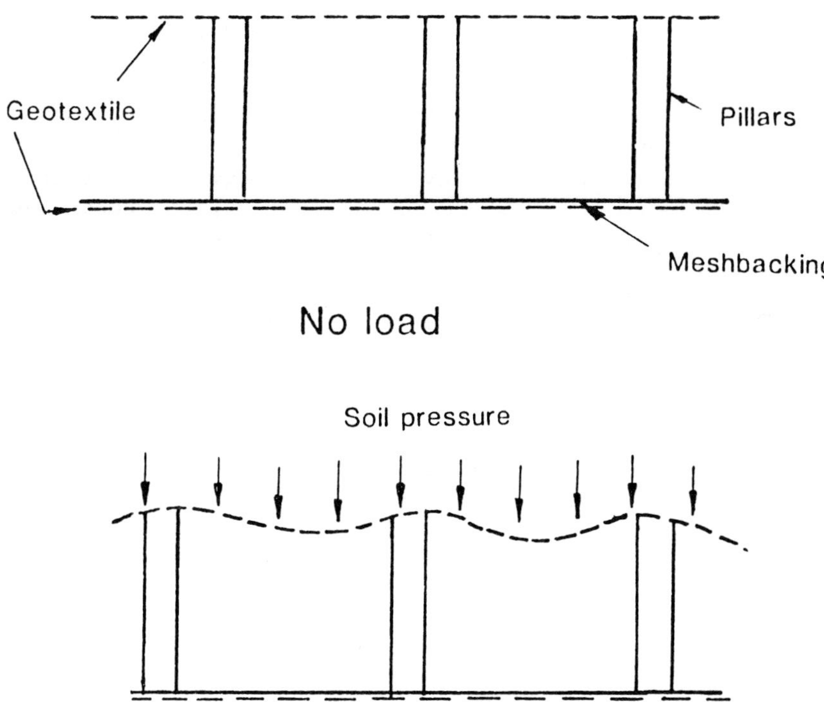

Fig. 1. Geocomposite showing intrusion by soil under load

Table 3. UK DTp filter design criteria

Soil Retention Criteria	Wovens and thin Non-wovens ≤2mm	Thick non-wovens >2mm
For soils with U_c between 1 and 5	O_{90}/d_{90}=1 for U_c=1 =3 for U_c=5	O_{90}/d_{90}=1 for U_c=1 =3 for U_c=5
For soils with U_c between 5 and 10 (Alternative Criteria)	$O_{90}/d_{90} < 3$ or $O_{90}/d_{90} < 1$	$O_{90}/d_{50} < 6$ or $O_{90}/d_{50} < 1.8$
Permeability Criteria (Note that Kg > K_s)	Kg > 10 K_s	Kg > 100 K_s

Murray and McGown [Ref. 2] suggest the criteria in Table 3.

For more severe applications in hydraulic structures, e.g. beneath revetments in tidal or navigated waterways, the soil retention and permeability criteria need to be more stringent. The more stringent criteria will ensure that piping through the geotextile does not take place and that rapid equalisation of water pressures can occur. The Dutch Rijkswaterstaat have the criteria shown in Table 4 [Ref. 3].

Numerous other criteria exist in geotextile literature and manufacturers' guides. The criteria are the subject of continued research both in Europe and in North America. At the 5th Geosynthetic Research Institute (GRI) Seminar at Drenel University (December 1991) Luettich *et al.* presented further guidelines in their paper [Ref. 4]. The approach presented is a 9 step procedure which considers the filter functions, survivability and other requirements.

The nine steps suggested are

Step 1: Define the Application Filter Requirements
Step 2: Define Boundary Conditions
Step 3: Determine Soil Retention Requirements
Step 4: Determine Geotextile Permeability Requirements
Step 5: Determine Anti-clogging Requirements
Step 6: Determine Survivability Requirements
Step 7: Determine Durability Requirements
Step 8: Miscellaneous Design Considerations
Step 9: Specify a Geotextile Filter

5. Design of geocomposites

The design of the core of a geocomposite drain must satisfy the following criteria.

(a) The core must resist the applied loads (normal and shear) without collapsing.
(b) Under sustained load the core must not reduce significantly in thickness (compression creep).
(c) The core must allow the expected water flow to reach the discharge point without the build up of fluid in the core.
(d) The core must support the geotextile filter.

At present there are only a few guidelines for the acceptance of geocomposites for particular uses. The Department of Transport at present restricts their use in highway works to road pavement sub grade drainage or groundwater control applications, their use as structural drains behind earth retaining structures and ground culverts is prohibited. Current reviews of existing practices are being undertaken by TRL and may result in the restrictions being lifted. Geocomposite drains have been used for structural drainage for at least 10 years on major and minor structures with no reported failures to date.

Table 4. Dutch Rijkwaterstaat filter design criteria

Situation	Soil Tightness Number and Requirement
Stationary	$D_{90}/d_{90} \leq 2$
Cyclic with natural filter	$D_{98}/d_{85} \leq 2$
Cyclic without a natural filter	
a) Extreme Conditions	$D_{98}/d_{15} \leq 1.5$
b) Normal Conditions	$D_{98}/d_{15} \leq 1.0$

6. Specification

Many specifications for geotextiles and geocomposites are written as 'Brand X or similar approved'. Specifications written in this way actually restrict the Contractor's choice to 'Brand X' unless the decision as to which geotextiles are 'similar' is referred to the design engineer. The Contractor has no information on which to make a comparison between products and the Engineer's Representative on site is often no better informed. The only way to give the Contractor freedom to choose from the products available and hence to obtain the lowest price for conforming product is to specify by quoting the required properties for the geotextiles to be incorporated in the works.

Once detailed design of the geotextile or geocomposite for the project is complete the specification based on the results can be written. Reference to manufacturer's brochures should be made at this point to ensure that geotextiles with the required properties can be obtained. It is unusual with the wide range of products available to be unable to find an acceptable product.

The geotextile filter should only be specified using those properties which are essential to its performance in the works. To ensure that the geotextile filter can be placed and will have sufficient strength and puncture resistance when in contact with the soils, minimum values of these parameters should also be specified.

The test procedures currently available (British Standards, International Standards (ISO) and those being developed for Europe by CEN), are all 'index' tests. The index tests are tests carried out on specimens of geotextile or geocomposite without the presence of soil. The actual field performance of geotextiles and geocomposites varies from the 'index' characteristics. Therefore for large projects where the performance of the geotextile or geocomposite is very important and sufficient relevant data on performance is not available then 'performance' tests, field or laboratory, should be specified before the geotextile or geocomposite is selected.

Standard specifications for geotextile filters are to be found in the DTp Specification for Highway Works (SHW) [Ref. 5] and the draft ICE Specification for the use of Geotextiles and Related Materials [Ref. 6].

For highway works the SHW [Ref. 5] requires the geotextile filters for fin drains to be specified using the following criteria.

(a) Tensile strength (BS 6906 Part 1 [Ref. 7]). (Min. 5 kN/m)
(b) CBR puncture resistance (BS 6906 Part 4 [Ref. 8]). (Min. 1200 N).
(c) Tear resistance (ASTM D 4533-85 [Ref. 9]). (Min. 200 N).
(d) Apparent opening size (AOS) O_{90} (BS 6906 Part 2 [Ref. 10]). (Appendix 5/4).
(e) Water flow normal to the plane (BS 6906 Part 3 [Ref. 11]). (Appendix 5/4).
(f) Breakthrough head (BS 6906 Part 3 [Ref. 11]). (Max. 50 mm).

The values of the first three parameters and the last are fixed in the specification, the water flow and the AOS are to be specified for each project by the Engineer following his design calculations based on the soils found on the site.

The SHW [Ref. 5] Clause 609 sets out the requirements for geotextile separators for general use; these are specified using the criteria

(a) Tensile strength (BS 6906 Part 1 [Ref. 7]). (Appendix 6/5).
(b) Water flow (BS 6906 Part 3 [Ref. 11]). (Minimum 10 $l/m^2/s$).
(c) Effective opening size (BS 6906 Part 2 [Ref. 10]). (0.1 to 0.3 mm).

The tensile strength required is to be determined for each project application in Appendix 6/5, while the other values are fixed for all highway works.

Engineers should be careful when presented with products which claim to meet the requirements of these two SHW Clauses as for each project the values entered in Appendices 5/4 and 6/5 must be checked against the certified properties of the geotextiles.

The draft ICE 'Specification for the use of Geotextiles and Related Materials' [Ref. 6] suggest that the following properties are specified for geotextiles used in Filtration.

Index properties

Mass per unit area
Nominal thickness/dimensions
Effective Opening Size
Percentage Open Area
Tensile Strength and Strain
Puncture Resistance
Tear Resistance
Permittivity

Performance properties

Long term thickness under service pressures
Water flow in soil
Clogging resistance
UV radiation resistance
Alkali resistance
Biological resistance

For Erosion Control, i.e. geotextiles beneath revetments, the two lists are extended to include

Index properties

Abrasion Resistance
Friction
Flexibility

Performance properties

Friction
Acid resistance

The extended list of properties which the ICE Specification [Ref. 6] suggests is a comprehensive selection of tests which would fully describe the geotextiles in use. However for most applications the shorter list of requirements used in the SHW will be adequate.

The final part of the specification is to detail the testing required to ensure that the geotextiles and geocomposites actually have the specified properties. Manufacturers 'glossy' brochures are often relied upon by site staff and others for certification. The values in these brochures are often average or typical values and cannot be taken as 'Quality Assurance' documents for the geotextiles used on site. Independent testing by properly equipped laboratories is an essential feature of any specification. The SHW [Ref. 5] requires that all geocomposite drains for use as 'fin drains' to Clause 514 have a current Certificate issued by the British Board of Agrement and that geotextiles complying with Clause 609 are sampled on site and tested by an 'approved laboratory to prove that the geotextile meets the specified criteria'. Few Engineers seem to be aware of the requirements set out in Clause 609 which were also present in the 6th Edition of the SHW [Ref. 12]. Clients and those responsible for Quality Assurance must ensure that the testing required by specifications is carried out to ensure that sub-standard materials are not used. The effects of sub-standard materials may not be immediately apparent and 'failures' may take many years to develop.

In the papers by Farrar and Myles [Refs 13 and 14] details of the proposals for a European Classification System are discussed. The implementation of a classification system with specific properties for grades within application functions, would enable Engineers to select a specific class of geotextile for a project application. The testing and certification which is part of the CEN 'CE' marking system of materials would enable Engineers to reduce their specifications and testing requirements to simple statements of compliance with relevant CEN Standards. However this ideal is some way in the future and detailed performance specifications will be required until the full implementation of CEN standards.

7. Conclusions

The properties of geotextiles, and how the method of manufacture and the polymers used affects them, need to be understood before design is undertaken.

Properties of the soils which will be in contact with the geotextiles must be obtained from ground investigation and laboratory tests.

The design of geotextile filters needs to be carried out using one of the sets of design guides appropriate to the application in the project.

'Brand X or similar approved' methods of specification must not be used.

Performance specifications should avoid using a multitude of properties which are not directly relevant to the application.

Manufacturers brochures must not be taken as 'certificates of compliance'. Proper independent testing of geotextiles and geocomposites is essential to ensure that products comply with the specification.

8. References

1. BS 6906 Part 7, British Standard Methods of Test for Geotextiles : *Determination of in plane water flow.*

2. MURRAY R. T. and MCGOWN A. *Ground Engineering Applications of Fin Drains for Highways.* TRL Application Guide No. 20, 1992.

3. BALKEMA A. A. *Geotextiles and Geomembranes in Civil Engineering*, Rotterdam, 1986.

4. LEUTTICH, GIROUD and BACHUS, Geotextile Filter Design Guide, 5th GRI Seminar Drexel University, 1991, to be published in *Geotextiles and Geomembranes.*

5. DEPARTMENT OF TRANSPORT (1991). *Manual of Contract Documents for Highwork Works, Volume 1, Specification for Highway Works*, HMSO.

6. INSTITUTION OF CIVIL ENGINEERS (1987). *Specifications for the use of geotextiles and related materials*, London, November.

7. BS 6906 Part 1 British Standard Methods of Test for Geotextiles : *Determination of the Tensile Properties using a wide width strip.*

8. BS 6906 Part 4, British Standard Methods of Test for Geotextiles : *Determination of puncture resistance (CBR Test).*

9. ASTM D4533-85 *Trapizoidal Tear Test for Geotextiles.* American Society for Testing of Materials, 1985.

10. BS 6906 Part 2, British Standard Methods of Test for Geotextiles : *Determination of the apparent Pore Size Distribution by Dry Sieving.*

11. BS 6906 Part 3, British Standard Methods of Test for Geotextiles : *Determination of Permeability by Water Flow.*

12. Department of Transport (1986) *Specification for Highway Works*, HMSO.

13. FARRAR D.M. Filter and Drainage Design for Highways. *Proceedings*

Geotextiles in Filtration and Drainage, Thomas Telford, London, 1992.
14. MYLES B. The Way Forward in Europe, *Proceeding Geotextiles in Filtration and Drainage*, Thomas Telford, London, 1992.

The hydraulic properties of geosynthetic products

I. SPENCE, Dundee Institute of Technology, UK

1. Introduction

The enhancement of soil properties by the inclusion of organic materials has been long established but it is only in recent decades that the use of man-made materials as a conduit within the soil has become an acceptable practice. It is not uncommon to see geo-drains being installed at the edge of carriageways to replace the more traditional french drains and it is a well established practice to wrap gravel drains in a fine geotextile to act as a filter and thus prolong the active life of the drain. In waste management the use of geosynthetic drains to replace the more traditional gravel layers which act as traps and conduits for leachate, gas and groundwater, is gaining in popularity, particularly as it can significantly increase the space available for the storage of waste. Despite its notoriety for ultra-conservatism, the construction industry, when it 'discovers' a new product, seems to place no bounds upon its ingenuity especially when it comes to applying the discovery to increasing the profit margin on a contract. The employment of materials based upon their past uses rather than upon their known properties and patterns of behaviour is often an acceptable practice in engineering where the factors of safety are such as to allow for reasonable variation within materials. In ground engineering, the variation in the properties of the soil can make any variation within the geosynthetic material seem insignificant and therefore hardly worth consideration. One might feel justified in considering the money spent on testing to be a waste and therefore a factor to be ignored when tendering for contracts, especially when using quality-assured factory-produced materials with very little variation between samples.

There are two principal reasons for testing materials. The tests can be performed in order to provide reliable data for use in the design process. The results of such tests are used to determine the allowable bearing pressure below foundations, the effort required to achieve an acceptable compaction

Geotextiles in filtration and drainage. Thomas Telford, London, 1993

of fill material or the size of conduit required to transport a given volume of water away from a carriageway. These tests are performance tests and should be designed to give the engineer a value related to the performance of the material under specified conditions. The tests must therefore be specified with sufficient flexibility to permit the in-situ conditions to be modelled (or simulated in some way), and yet with adequate restraints to ensure uniformity of results from test to test both within and between laboratories. The other family of tests are the index tests which may be performed to determine the compliance of a material with a given specification. It is crucial that these tests are unambiguously defined in terms of their apparatus and procedure in order to reduce the experimental variation due to the test to a minimum and emphasise the variation due to the material under test.

It is therefore important that tests should be specified in such a way that the recipient of the results is made fully aware of the use to which the values may be employed. A test which is designed to measure the flow of fluid along the plane of a geo-drain which is confined between rigid platens, may not give a value which may be used to calculate the flow expected when the same material is placed in the soil. The in-situ flow may be influenced by the confining pressure on the material, the soil type, the intrusion of the fabric cover into the core, the properties of the fluid to be transported the hydraulic gradient etc., all of which are site specific variables of importance to the engineer who is undertaking the design phase of the project. Is it realistic however, to expect a specification test to take all of these factors into account? It is therefore important when considering the testing of materials to realise that there may be significant differences between the tests required to permit the engineer to undertake a safe and cost effective design and those required to ensure compliance with a given specification.

2. Tests for the hydraulic properties of geosynthetic materials

The properties which must be considered when assessing the performance of geosynthetic materials for their use as hydraulic media are those which permit, or restrict, the flow of a fluid either along or through the product. These are, the size of the openings in the material, the ability of fluids to flow along the plane of the material (Transmissivity) and the ability of fluids to flow through the material (Permittivity).

2.1. Determination of the pore size

The three tests which are used for the determination of the pore size of geotextiles all employ the material which is under test as a sieve. This is the only thing which they all have in common. Two of the tests employ vibration to 'encourage' the test particles to pass through the fabric, although in one case the sample and the test particles are dry and in the other the test particles are washed through the fabric. Even this is further complicated by the fact that the dry sieving method specifies single-sized glass ballotini as the test particles and the wet sieving specifies well-graded sand. The third method also employs a well-graded granular material as the test particles but effects the sieving by oscillating the fabric with the test particles on its upper surface vertically in a tank of water.

With such diversity in the test method it should not be surprising that the tests do not specify the pore size of the fabric in the same terms. The dry sieving determines the Apparent Pore Size (O_{90}) which is defined as being a measure of the pore size of a geotextile determined by the passage of dry spherical solid glass beads through the geotextile under specified conditions, with O_{90} being the bead size, in microns, for which 90% of the beads are retained on or within the geotextile (BS 6906: Part 2). With the two wet methods it is the Effective Pore Size which is determined by comparison between the grading curves of soil samples above and below the fabric after a specified period of sieving. The Effective Pore Size is the mean particle size below which 10% of the sample passes through the geotextile.

Despite the obvious diversity within the test methods it has been shown that the values obtained from the dry and wet sieving methods give acceptably similar results, and that although the hydrodynamic method gives a higher value this is probably due to the difference in the soil samples utilised in the test method (Van der Sluys and Dierickx 1990). If the test is to be employed as an index test for specification purposes, then it should be standardised to a short duration simple sieving method using single-sized fractions which can be rigorously specified and undertaken with relative ease and minimum cost. Where the test is to be performed for specific purposes such as assessing the suitability of a textile for use in erosion control or as a shutter liner, then test methods should exist of sufficient flexibility to enable the test to determine reliably the performance of a material under specific conditions. Such tests, however, require an understanding of how the materials behave in the given conditions.

2.2. *Water flow through a geotextile*

The measurement of the flow of water normal to the plane of a geotextile would initially appear to offer little scope for test permutations. All that is required is to permit water to flow through a sample of material at a measured flow rate and determine the loss of head across the geotextile. A relatively simple test is described in BS 6906: Part 3: 1989. The test involves mounting a sample of geotextile in a permeameter and adjusting the flow until a constant head loss across the fabric of 100 is achieved. Once the head

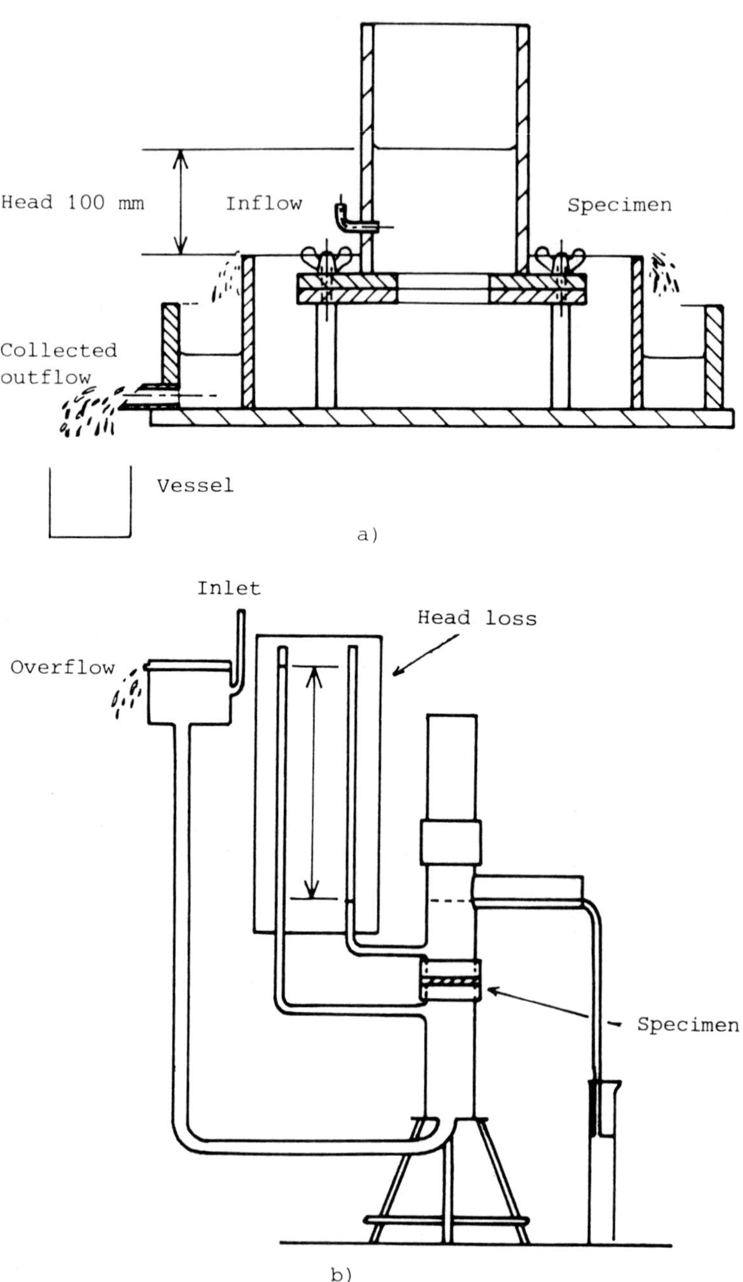

Fig. 1. Apparatus to determine the permittivity of a geotextile: (a) open system; (b) closed system

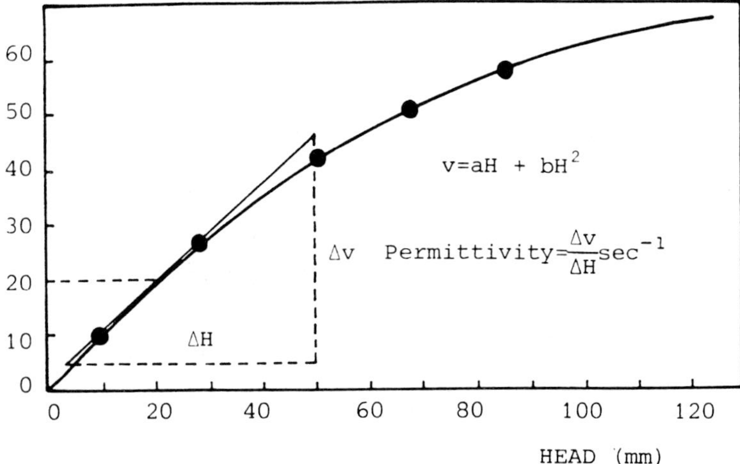

Fig. 2. Relationship between head and velocity for the flow of water through a geotextile

loss has stabilised at this value the flow of water is determined. This is expressed in L/sm^2 with the values being corrected for temperature to a standard value of 20°C.

This test is, however, not without its problems. There is no standard design for the apparatus and so the test may be carried out under open or closed conditions (Fig. 1). The test may be performed with the sample in the horizontal plane with the flow of water moving vertically upwards or downwards through the specimen. In some double cell permeameters the sample may be mounted in a vertical plane and the water flow is then horizontal. Although the test is normally carried out on a single thickness of material, it is possible to use multiple thicknesses of fabric. Thus it is obvious that even a relatively simple test is not immune from unnecessary variations.

The test as described may be used to produce one of two simple index values. The method described above produces a flow index: the flow of water in mm/s to produce a loss of 100 mm head across the sample. Some specifications ask for the production of a head index: the head loss which will be produced by a flow of 20 mm/s through the specimen. Recent proposals have suggested yet another value to assess the flow normal to the plane of a geotextile. This requires the apparatus to be set up to give a flow velocity of approximately 60 mm/s. When the flow has stabilised at this value the head is recorded and the actual flow measured. This is repeated four times with a range of flows evenly spaced between 0 and 60 mm/s. The velocities, corrected to 20°C are then plotted against the head and the best fit equation of the form $V = aH = bH^2$ calculated. The value which would then be used as the index value would be the gradient of the tangent at a flow of 20 mm/s (Fig. 2).

2.3. *Water flow along a geotextile*

The flow of water parallel to the plane of a geotextile is termed its Transmissivity. The test to determine this is BS 6906: Part 7: 1990: The determination of the in plane water flow. This test is almost identical to that described in ASTM D4716-87. The test apparatus (Fig. 3) permits water to flow along the plane of a geotextile under a constant hydraulic gradient of either 0.1 or 1. The flow is measured and the transmissivity is calculated from the equation

$$T = (QL)/(WH)$$

where T is the hydraulic transmissivity (in m^2/s), Q is the average quantity of fluid discharged per unit time (m^3/s), L is the length of the specimen (in m), W is the width of the specimen (in m), and H is the loss in total head across the test specimen (in m).

This value for transmissivity is dependent upon the hydraulic gradient and therefore must be quoted along with the corresponding value of the hydraulic gradient.

This test is relatively simple to perform but experience has shown that the results are not always reliable. For low-flow materials it is difficult to collect sufficient quantities of water over a reasonable period of time and so the results may be less reliable. Test conditions with high flow geosynthetic

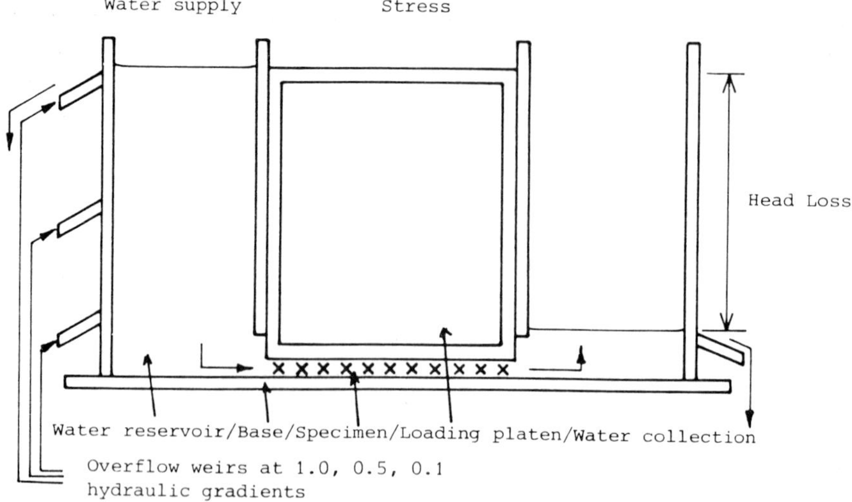

Fig. 3. Apparatus to determine the transmissivity of a geotextile drain

drains are frequently such that problems may occur in providing a sufficiently high flow of water into the apparatus to maintain the constant head without drawing air into the test specimen. Much work still requires to be carried out on the transmissivity test to improve its reliability. The design of the equipment and the interpretation of the results must be carefully investigated in order to produce a simple reliable index test from which it may be possible to develop a competent performance test

2.4. *Why test geotextiles?*

Given the above complexity of test methods and interpretation of results, the construction industry may well ask why it should waste money on the testing of geotextiles if the results can be so variable. If a contractor, by virtue of previous experience, may select a test house employing a particular test method which complies with the acknowledged standards and gives the values required to comply with a contract specification then there is little point in asking for materials to be tested. When tests for water flow through a geotextile carried out in 15 different testing centres give a value of 11.5 L/m/s with a standard deviation of 5 (i.e. a coefficient of variation of 43.6%) what reliance can the engineers put on the values when using them in their design calculations?

Before specifying tests for the hydraulic properties of geosynthetic materials the question must be addressed as to why the tests are being performed. If these are to be index tests which will be set as quality assurance standards on the materials supplied to a contract then they should be uncomplicated tests on the basic properties of the materials. They should be understood as such and not expected to give the engineer values for use in design. They should be tests which are as readily available to the contractor as they are to the manufacturer and should not require large expenditures of money to provide routine test certificates. In such tests the apparatus should be rigorously specified to avoid variation due to test equipment and the test method should be sufficiently detailed in terms of routines that they may be performed with competence on a regular basis, thus reducing error derived from operator short cuts and slipshod practice.

When values are required for design purposes testing moves into a totally different sphere. The engineers require values which will enable them to produce a safe cost effective design. This requires much more information on the in service behaviour of the materials. The use of geotextiles as filters has been widely researched and the engineer, when selecting a filter fabric, gets information on the efficiency of different types of fabric in building up a filter cake. This will depend upon the types of soil in which the filter is to be employed, the hydraulic conditions etc. If these are known and the required performance criteria for the fabric are understood then the engineer should be able to select a material specification in terms of index properties for the tender documents. Unfortunately it is at this point that the problems of such a two-fold approach to testing arise. What is the relationship, if any, between the simple index test for quality assurance and the performance of the material in-situ?

The simple answer is to state that the index values are for quality assurance purposes only and should not be taken as indicators of in service performance. This may satisfy the purists in quality assurance but would be of little value to the users of the products. Manufacturers would probably find themselves producing two sets of values; those for quality assurance and hence specification purposes and those for use in the design calculations. This problem should not be insurmountable. If a set of index tests could be established and accepted then work could be carried out, using these tests as a baseline, to identify the performance criteria related to this base, and therefore once an engineer knows the requirements in terms of the site performance of the materials the tender specification could then be more meaningfully selected.

3. Index tests for the hydraulic properties of geosynthetic products

Geosynthetic materials used as filters or drains must be capable of transmitting the fluid for which they have been specified. This means that the fluid must be able to flow into the material and either pass through it or flow along its plane. Filters must also be capable of forming the base upon which a filter cake may develop. The material specified must therefore be capable of permitting the passage of a fluid whilst, in the case of a filter, impeding the passage of solid particles. The requirements for a set of index tests is therefore to indicate the transmissivity, permittivity and pore size of the material.

3.1. Transmissivity test

The apparatus used in the British and American standard test for fluid flow parallel to the plane of a geotextile permits the material to be confined under a variety of normal loads with the hydraulic gradient across the specimen being either 0.1, 0.5 or 1.0. As initial work has shown that problems arise when attempting to use high flows through the equipment, it would be prudent for an index test to use lower flow conditions; i.e. a hydraulic gradient of 0.1. In order to minimise the influence of creep of the material a relatively light confining load should be applied to the specimen and maintained for a specified time. A load of 25 kPa should be sufficient to seal the sample in the apparatus. In order to permit the flow to stabilise, the water should flow at a constant head for a minimum of ten minutes prior to measuring the flow through the material. The flow should be measured by taking at least 5 independent samples of water from the outflow weir with a minimum volume per sample of 0.5 litre and a minimum time of 5 seconds.

The index value would then be the flow in litres per second to pass through a sample of material 200 mm in the cross flow direction and 300 mm in the direction of flow under a confining pressure of 25 kPa, and a hydraulic gradient across the specimen of 0.1, after the water has been allowed to flow for a period of 10 minutes. The index should be taken as the mean of at least 5 values of flow calculated from the collection of at least 5 different volumes of water. No sample should be less than 0.5 litres and have taken less than 5 seconds to collect. The values should be corrected to a temperature of 20°C.

3.2. *Permittivity test*

The present British Standard test , BS 6906: Part3: 1989 has been subject to a significant amount of work prior to the preparation of the standard and would form a good base for an index value if it were to be more closely specified as to apparatus and specimen size. The problems associated with this test arise primarily with achieving the required head of 100 mm for the more permeable geotextiles when a sample of 100 mm diameter is used. The index value should therefore be based on a smaller sample diameter (50 mm) with a lower head loss across the sample (50 mm). In order to maintain consistency with the transmissivity index the apparatus should be such that the water levels are determined by an open weir system rather than a fully confined permeameter and the requirements for settling time, flow measurement and temperature should be kept the same.

The index value for the flow of water through a geotextile would then be the flow in litres per second to flow through a 50 mm diameter single thickness sample of geotextile with a head loss of 50 mm across the sample after the water has been allowed to flow for a period of 10 minutes. The index should be taken as the mean of at least 5 values of flow calculated from the collection of at least 5 different volumes of water. No sample should be less than 0.5 litres and have taken less than 5 seconds to collect. The values should be corrected to a temperature of 20°C.

3.3. *Pore size test*

In order to compare the pore size of different geotextiles there should be one, well-defined, simple sieving method of short duration. As this is an index test the apparatus should be as uncomplicated as possible and so a dry sieving test which avoids the necessity of complex spray heads, water flow control and catchment systems should be chosen. The sample size can make a significant difference in the overall costing of a test programme especially if a high proportion of the material supplied to the site is to be tested, and so the size should be restricted to the minimum size which will give reproducible results, in this case using a sample of 250 mm diameter in a 200 mm diameter sieve. The choice of test particles between sand or ballotini has been shown by Van der Sluys and Dierickx (1990) not to alter significantly the pore size determination, and in order to make standardisation more easily achievable single-sized spherical ballotini should be used as the test particles. The sieve shaker should be capable of a constant vertical amplitude of vibration of between 0.5 and 0.75 mm. The test method as specified in BS 6906: Part2: 1989 with the above modifications would provide an adequate index test for the determination of the pore size of geotextiles.

4. Conclusions

In order to assess the hydraulic properties of geosynthetic materials it is necessary to determine the pore size of the material and its ability to permit the passage of fluids both normal and parallel to the plane of the material. The range of tests currently used to determine these properties is such that it is not always possible to determine to compare one set of values with another. This is particularly important when the test values are being used as a method of specifying the materials within a contract document or when attempting to meet the requirements of quality assurance testing.

It is therefore necessary to produce a set of index values which may be used in the specification of geosynthetic products. These should be closely specified tests to determine index values related to the pore size, transmissivity and permittivity of the materials. Once such a set of tests is established, research will be required to relate these index values to the in-service behaviour of the materials, and hence develop performance tests which will provide the engineer with reliable data on the hydraulic properties of geosynthetic materials.

5. References

ASTM d4716-87. *Test Method for Constant Head Hydraulic Transmissivity (In Plane Flow) of Geotextiles and Geotextile Related Products*. American Society for Testing Materials

BS6906: 1989. *Methods of Test for Geotextiles.* British Standards Institution

VAN der SLUYS and DIERICKX (1990). Comparative study of different porometry determination methods for geotextiles. *Geotextiles and Geomembranes*, 9, pp 183 - 198

Discussion

Reporter: J. Dixon, Netlon Ltd, UK

C. LAWSON, Exxon Chemical Geopolymers

Test method variation must be less than product variation otherwise one will only measure test variation not the product variation. Unfortunately the situation exists with CEN European tests and the American ASTM tests that pore size, transmissivity and permeativity tests have a variation in test method similar to the product variation. The test variation causes fundamental statistical problems, particularly for control tests. Fig. 1 shows the relationship between the statistical interpretation and the specification.

In carrying out the pore size tests there are fundamental problems with equipment variations and operator variations. BS 6906 : Part 2 states that the test results can have a variability of ±20%. However, engineers seem to ignore the problems when writing specifications. On one particular job in UK, an Engineer had a specification based on a mean value of 100 μm (the family mean) but on site the Engineer's Representative interpreted the specification that the tests were required to have a sample mean of 100 μm, a different set of requirements; the sample mean was 80 μm and the material was rejected as being below spec. value. Two other samples from the same roll were tested at another laboratory which gave different results, which confirmed acceptance, as the family mean was greater than 100 μm.

Requests: Specifications must be written on a proper logical statistical basis. Specifications must be interpreted on the same basis as they are written. Index tests can only be used for control purposes if the test variability is less than the product variability.

CHAIRMAN

Does your company and others quote variance and family mean?

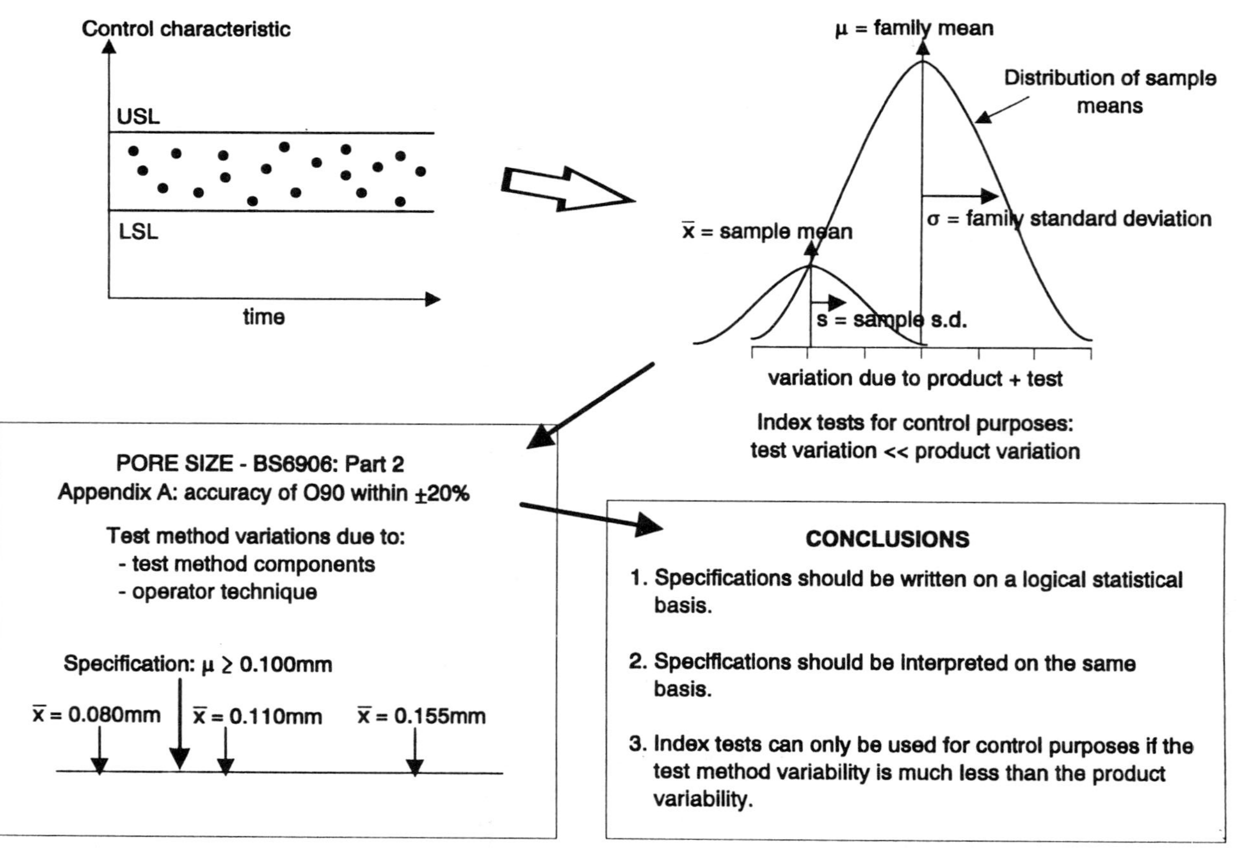
Control characteristic
USL
LSL
time
μ = family mean
Distribution of sample means
σ = family standard deviation
x̄ = sample mean
s = sample s.d.
variation due to product + test
Index tests for control purposes:
test variation << product variation
PORE SIZE - BS6906: Part 2
Appendix A: accuracy of O90 within ±20%
Test method variations due to:
- test method components
- operator technique
Specification: μ ≥ 0.100mm
x̄ = 0.080mm
x̄ = 0.110mm
x̄ = 0.155mm
CONCLUSIONS
1. Specifications should be written on a logical statistical basis.
2. Specifications should be interpreted on the same basis.
3. Index tests can only be used for control purposes if the test method variability is much less than the product variability.

Fig. 1

C. LAWSON, Exxon Chemical Geopolymers
If people ask we will supply the values but brochures can only contain a limited amount of data.

CHAIRMAN
As these are important perhaps they should be stated.

D. FARRAR, TRL
It is a real problem for specification but it is not necessarily a design problem. If the designer is able to predict the flow to ±50% he will probably be satisfied. Therefore a large range in test results does not cause a problem in design.

CHAIRMAN
It may be difficult to define the test so that laboratories can carry out the tests properly.

These discussions could have been found in a different context 30 or 40 years ago when researchers were looking at the permeability of peats. It eventually became understood that the effective stresses acting on the material were a fundamental part of the test. Some of the problems presented by Iain Spence could be explained by the low effective stresses in the hydraulic tests. Effective stress controls porosity and porosity affects permeability.

B. MYLES, Consultant
If this were true then the effects would be seen more on thick compressible fabrics; in actual fact the effect is seen in both thick and thin fabrics. The change in behaviour can be explained by the change from laminar to turbulent flow. I agree that effective stress state could contribute but feel that it is only a small part.

CHAIRMAN
This is interesting, but in gravels the flow usually only becomes turbulent at about 10 mm particle size. Are we suggesting that this happens earlier in geotextiles?

C. TUXFORD, Euro Erosion Engineering
Use of index tests can make life difficult for non-UK manufacturers in the UK market, as engineers/specifiers insist on British Standards tests. Why should UK contracts insist on BS tests, and should not DIN or other national tests also be accepted?

D. MANN, Applied Geology
A potential problem is that several of the design methods are empirical and

related to a particular test method; therefore we must be careful when accepting other test methods, without reviewing the design.

B. MYLES, Consultant
The only place this can arise is in the pore size test, particularly in Germany where the wet sieving test gives a pore size different to other methods. Generally the variation in pore size is usually only 10-15% and hence different index tests can be used.

CHAIRMAN
There are two issues here: the difference between test methods and calculations carried out; whether any of this relates to reality (Prof. Rowe's presentation included a geotextile which should have failed, based on design, but in laboratory tests the geotextile performed well and failure was actually caused by poor workmanship).

If this is looked at in effective stress terms, when the soil is in a state of low effective stress the soil particles can be easily moved. However, if the correct effective stresses are applied the soil particles are jammed together and will be less mobile. So, do the Index Tests represent reality? If not they are probably very conservative.

B. MYLES, Consultant
I agree, but when the different design methods are considered there are variations of factors up to 10 between the results. One can use any design method; Terry Ingold can show large variations between the different design methods, larger than differences between test methods.

S. CORBET, G. Maunsell & Partners
There is a difference between Index and Performance Tests. The former should only be used for quality control purposes, but in the absence of other data the results of Index Tests are being used with various factors to suit local conditions. The Index Tests have been designed (BS, ASTM, ISO, DIN, CEN etc.) to give an indication of the basic properties of geotextiles. However as engineers we wish to carry out analytical design, but without proper Performance Tests which model real site conditions (the next work programme for CEN TC 189) the data for proper analytical design are not available.

I. SPENCE, Dundee University
O_{90}/d_{10} values are being quoted as varying by factors of 5 or 10. We must be careful: these are not the same thing. Much more work is needed to identify what we are measuring.

B. MYLES, Consultant
In the ISO round robin tests which involved 18 laboratories, 4 geotextiles and 3 methods, and took over 1 year to complete, the results were remarkably similar. I feel that the problems are elsewhere and that stress levels are important when looking at flow and pore size characteristics. The factors applied to index tests in design are purely arbitrary and must be treated with caution.

K. ROWE, University of Western Ontario
As there is a large variation in the design recommendations we need to recognise that when there is a failure different 'experts' can be found to support both sides. We must be careful that we do not hinder progress with unnecessarily over-conservative design.

P. RANKILOR, Manstock Consultants
Equations that we have been looking at are inequalities and care must be taken in selecting the limit values. How do we look at the lower limit part of the inequality? The state of compaction and the hydraulic gradient in the soil are important factors. One of the limiting factors on the geotextile may be the hydraulic pressure behind the geotextile. In particular conditions the lower limit in the inequality may need to be reduced significantly to maintain equilibrium in the soil structure.

CHAIRMAN
Hydraulic gradient is important; other researchers have put a lot of effort into looking at these problems and Prof. Rowe made the point that flow concentration into drains was a poor design detail: blankets which spread the flow and hence reduce hydraulic gradients are better.

D. AYRES, Consultant
When looking at many granular blankets over clay soils, I have observed that the interface is not a clear line but is spread over 20/50 mm. We are now introducing a thin geotextile filter only 1 or 2 mm thick instead of a sand filter 150 mm thick; the transition zone cannot be develop.

D. FARRAR, TRL
We have also observed the formation of transition zones in soils in contact with geotextiles and suspect that there may be the potential for failure surfaces to develop. It is therefore not a simple boundary but a complex structure.

CHAIRMAN
We see the development of a structure more complex than designers wish,

but these manufactured materials clearly have a place. We need Index Tests to see what we are looking at and that there is scope for designers to use those materials in ingenious ways. The progress of standardisation and specification seems to be well underway.

Geotextiles in long-term use: synopsis of the K & O report

H. WICHERN, K & O Committee, The Netherlands

1. Introduction

The use of geotextiles in hydraulic engineering has greatly increased in recent years. Geotextiles can be used in the simplest of drainage channels to multi-functional structures like the Eastern-Scheldt Closure Dam in Holland.

Geotextiles need to retain soil particles beneath the slope and be permeable enough to prevent the build-up of hydrostatic pressure.

In Holland, over the last 25-30 years much research has been carried out on the properties and the behaviour of geotextiles used as filters in civil engineering. Until the late 70s, however, no research on geotextiles that had an entire 'life' as part of a slope protection, had been initiated, thus the research of 'de Nederlandse Vereniging Kust- en Oeverwerken', during 1980-82 can be considered unique. The conclusions of the research are as follows.

The geotextiles considered in the research, all wovens, are between 10 and 15 years old. Over this period the influence of time on the physical properties is minimal. In cases where the properties have changed unfavourably, the reason was probably extraordinary loads during construction and service.

The technical and chemical properties of the woven fabrics are not influenced by conditions, either above or below water, if the material is exposed. It can be shown that soils containing iron (and oxide) reduce the life of Polypropylene geotextiles.

Reduction of permeability of the wovens as a result of blocking and clogging is found in most cases, however not to an alarming extent. Clogging did not occur in the case of an open construction subject to wave attack. Soils containing iron create a considerable amount of 'pollution' of the geotextile and can, due to hydraulic pressure, lead to failure of the construction.

Changes in the constitution of the subsoil mostly take place in a very thin layer (approximately 5mm) directly underneath the geotextile filter. The

Geotextiles in filtration and drainage. Thomas Telford, London, 1993

change of permeability in this layer and the subsoil are not directly related. The behaviour of the upper layer is influenced by the geotextiles.

The wovens considered in this programme are from slope protection systems along rivers and shipping canals. For these conditions, the geotextiles have to be more open, in all cases a factor 2 times the soil tightness requirements suggested by Ogink (Ref. 1).

The research was given to the Technical Committee of the Nederlandse Vereniging Kust- en Oeverwerken. This committee appointed a working group to advise the project manager Ir. R. Veldhuizen van Zanten. Prof.dr.ir. E.W. Bijker (Science), Ing. T. ten Bruggencate (Manager K & O), Ir. A.G.M. Groothuizen (Contractor), Ing. K.A.G. Mouw (Government), Ing. H.A.M. Wichern (Geotextile suppliers). In sample taking there was the assistance of G. de Keizer (1980) and E. Hakkenes (1981). At the request of the Technical Committee of K & O, at the start of the research an investigation of possible sample locations was carried out by A. Voorberg (Rijkswaterstaat). The actual research was done by several Laboratories: Delft Soil Mechanics Laboratory, who also made available the equipment for soil sampling, Vezelinstituut TNO, Kunststoffen en Rubber- instituut TNO, Delft Hydraulics. The results of the Vezelinstituut TNO have been evaluated by Prof. ir. K van Harten and Ir. A.H.J. Nijhof of Technical University, Delft.

2. Purpose and background of the research

The geotextile research at the Nederlandse Vereniging Kust- en Oeverwerken (K & O) has been carried out for 10 years. The technical committee in charge of the research on the orders of the 'Vereniging' started in the early seventies to study slope protection. The aim was standardization of the design of slope protection under various conditions. Standardisation should enable small contracting firms to tender for works involving coast and slope protection, particularly where alternative solutions are allowed or when the project details are left to the Contractor's own choice, subject to compliance with the specification.

Rijkswaterstaat, the municipality of Rotterdam and Delft Hydraulics had also participated in the work of the committee. For example in 1974 a large test was carried out in the Hartelkanaal in Rotterdam Port where the effect of water pressure under riprap was measured. The work programme gave answers but also prompted many questions. A lot of work needed to be done with various subsoils, under various loads and requirements for filters looking at soil tightness and permeability.

The technical committee appointed two sub-committees, one for bituminous and one for non-bituminous materials. In 1977 both sub-committees reported proposals for a number of standard details. In many of the standard details geotextiles were included.

In 1977 the first IGS conference took place in Paris and 'Geotextiles as Filter construction in Hydraulic Engineering' was discussed. However the Technical Committee felt that this conference showed that there was a distinct lack of knowledge and experience.

In June 1977, the technical committee, wanting to bridge the two events, appointed a working group and instructed them to produce guidelines for a research into the behaviour of geotextiles in real situations.

The working group formulated the geotextile research to be carried out as follows: 'Acquiring knowledge of the behaviour of geotextiles in slope protection as a function of time. The behaviour of geotextiles as a function of time is decided by the rate of ageing, clogging and blocking of the filter. In combination with earlier and recent research, a list of demands and standards for geotextiles should be produced. Also design standards for filter applications in hydraulic engineering should be produced as an extra to experimentally based design criteria.'

3. Geotextiles

In the original report (Ref. 2) there is a large chapter dedicated to the make up of granular filters, descriptions of polymers and their properties, fibres, yarns, wovens, nonwovens, etc. Now in 1992 this information is generally understood or can be read elsewhere.

4. Original properties of the geotextiles

One of the first conclusions of a pre-study was that it would be absolutely useless to take samples from a location when the original properties of the geotextiles used were not known. For that reason the pre-selection of test sites was focused on sites where the geotextile history was known, including the supplier, a record of site conditions, the wave attack, water level variations, chemical conditions etc. In reality however it turned out that knowledge about the geotextile supplied, was no guarantee that the original facts could be provided. There were various reasons for this. On the test site a different product to that indicated in the tender document had been used. Although similar to a certain degree, another quality was supplied, mostly for price reasons. In those pioneering days the standards were not exactly formulated and although it did not make much difference for the project itself, it did make the research difficult. Another complication was that products have been changed by manufacturers, but trade name or grade have not; in some cases the suppliers have changed the yarn type.

5. Preparation for the research

5.1. Set-up

Details like pre-study and financing the research will not be discussed here, but it is necessary to understand the reason for the large number of samples, taken from each sampling site.

At every selected location there are 2 possible sampling places

(a) as close to the waterlevel as possible
(b) a few metres above the waterlevel.

This produces a number of possible combinations shown below.

5.2. Number of samples

In practice there are many varieties of filter constructions. These varieties are combinations of the following.

Filter Type: A Mesh, B Tape Fabric, C Heavy fabric, D Non-woven.
Raw Material: A Polypropylene, B Polyethylene, C Polyamide, D Polyester.
Soil Properties: A Fine, D50 < 50 micron; B Middle, 50 micron < D50 < 150 micron.
C Coarse, D50 < 150 micron.
Location: A Under water, B At waterlevel, C Above water.
Water quality: A Fresh, clean; B Saline, clean; C Fresh, polluted; D Saline, polluted.
External Hydraulic conditions, e.g. calm or rough.
Internal Hydraulic conditions, e.g. perpendicular flow, parallel or both.
Exposure time.

Using the first 7 above mentioned items gives over 4300 theoretical possibilities. The research was reduced for practical reasons to

(a) Type of polymer: Polyolefine or other
(b) Soil Properties: Fine, D50 < 100 micron; or normal, D50 > 100 micron.
(c) Place: Above water level or on, or when possible, below water level.
(d) Water quality: Fresh or salt water.
(e) External: Calm or rough hydraulic conditions.
(f) Internal: Perpendicular or parallel flow.
(g) Time: Only samples older than 7 years.

Only 3 underwater samples were taken since it was generally considered to be impossible to obtain undisturbed soil samples. The 3 samples were taken when there was the opportunity to take advantage of temporarily low water levels (weirs, tide etc.).

For (e) calm is normally, Hs < 0.5m, V < 2.5m/sec., and rough Hs >0.5m,

V >2.5m/sec., where Hs is significant wave height and V is current velocity.
For (f) perpendicular is basically outflowing groundwater. Parallel is wave run-up, heavy rain, etc. A combination will occur in most places.
The combinations restrict the number of locations to 32, namely:

2 (1a + b), times 2 (2a + b), times 2 (3a + b), etc.

Not all combinations are of great interest however. Dense soil and outflowing water was one such case. The locations were rechecked and the final selection was as given in Table 1. Also at sites where there was a chance of interesting results, less important sites were ignored. The final 32 comprised 8 sites above waterlevel and 24 sites at the waterline (see Table 2).

6. Selection of sample sites

As there were 2 years allowed for the research, attempts were made to profit from experiences gained in the first year. Another point was the existence of maintenance or rebuilding plans. Therefore the Hartelkanaal in Rotterdam seaport was chosen for study during the first year. Additional reasons for selecting the Hartelkanaal in the City of Rotterdam were involved in the research: it was a relatively old site (1968), the hydraulic conditions (rough) and subsoil (fine) were interesting. These were the considerations for 3 of 12 first year test sites. All tidal tests were postponed until 1981, but other inland shipping canal sites completed the list.

The result of tests in 1980 produced recommendations for the sites to be studied in 1981, additional to the previous criteria. These preliminary results/facts/recommendations were

(a) no significant differences were found between wovens coming from the water level zone or the dry zone. Therefore at each site it was sufficient to take a sample from the water level zone.
(b) so far no PA (other) wovens had been taken in account and also no heavy fabrics (over 600 g/m^2) had been sampled.
(c) attempts were made to find sites where problems had been noted in the past.

Contrary to the recommendations in 1981 a few tests were taken from above as well as at the waterlevel at a few sites where heavy wave attack could be an influence on the geotextile performance.

6.1. Evaluation

Table 2 shows the selected sites used for this research. In some cases it was possible to take samples from below the watertable despite the original

Table 1. Final selected sites

	Material		Subsoil		Location		Water-quality		Hydraulic conditions		Current		Result
	PP / PE/	/PA	fine/	/normal	above/	/below	fresh/	/salt	normal/	/rough	perp./	/par.	
	1a	1b	2a	2b	3a	3b	4a	4b	5a	5b	6a	6b	
a	x	x	x				x		x		x		0
b		x	x				x		x		x		0
c	x			x		o	x		x		x		000
d		x		x		o	x		x		x		?
e	x		x					x	x		x		0
f		x	x					x	x		x		0
g	x		x				x			x	x		0
h		x	x				x			x	x		0
i	x			x		o	x			x	x		00
j		x		x		o	x			x	x		00
k	x		x			o	x		x			x	0000
l		x	x			o	x		x			x	?
m	x			x	o	o	x		x			x	0000
n		x		x	o	o	x		x			x	?
o	x		x			o		x	x			x	0000
p		x	x			o		x	x			x	?
q	x		x					x		x	x		0
r		x	x					x		x	x		0
s	x		x			o		x		x		x	0000
t		x	x			o		x		x		x	00
u	x			x		o		x	x		x		0000
v		x		x		o		x	x		x		?
w	x			x		o		x		x	x		0000
x		x		x		o		x		x	x		?
y	x			x	o	o		x		x		x	0000
z		x		x	o	o		x		x		x	00

	Material		Subsoil		Location		Water-quality		Hydraulic conditions		Current		Result
	PP / PE/	/PA	fine/	/normal	above/	/below	fresh/	/salt	normal/	/rough	perp./	/par.	
	1a	1b	2a	2b	3a	3b	4a	4b	5a	5b	6a	6b	
a'	x		x			o	x			x		x	0000
b'		x	x			o	x			x		x	?
c'	x			x	o	o	x			x		x	?
d'		x		x	o	o	x			x		x	?
e'	x			x	o	o		x	x			x	0000
f'		x		x	o	o		x	x			x	?

Notes

1. Location column. o = Sets of circumstances considered worth investigating.
2. Results column, (amount of information expected from this location)
   - oooo = large.
   - ooo = reasonable.
   - oo = average.
   - o = low.
   - ? = will it be possible to find such a set of circumstances?

Table 2. Conditions at the sites

Nr. Year	Location	Site Mat.		Subsoil	Water-quality	Hydr. circum	Current
	1980 research:						
1 '69	Noordzeekan.V	Abv	PP	normal	fresh	rough	parallel
2 '69	Noordzeekan.V	WL	PP	fine	fresh	rough	parallel
3 '74	Noordzeekan.B	Abv	PP	fine	fresh	rough	parallel
4 '74	Noordzeekan.B	WL	PP	fine	fresh	rough	parallel
5 '77	Maas-Waalkan.	Abv	PE	fine	fresh	rough	perp+par
6 '77	Maas-Waalkan.	WL	PE	fine	fresh	rough	perp+par
7 '70	Heumen kan.	Abv	PE	normal	fresh	rough	perp+par
8 '70	Heumen kan.	WL	PE	normal	fresh	rough	perp+par
9 '71	Wilh.kan.sluice	WL	PE	fine	fresh	calm/r	parallel
10 '70	Wilh.kan.out	WL	PE	fine	fresh	calm/r	parallel
11 '72	Z'sluis lake	Abv	PE	normal	fresh	calm	perp+par
12 '72	Z'sluis lake	WL	PE	normal	fresh	calm	perp+par
13 '68	Hartelkan.Vw	Abv	PP	fine	salt	rough	perp+par
14 '68	Hartelkan.Vw	WL	PP	fine	salt	rough	perp+par
15 '69	Hartelkan.SL-w	Abv	PP	fine	salt	calm/r	perp+par
16 '69	Hartelkan.SL-w	WL	PP	fine	salt	calm/r	perp+par
17 '69	Hartelkan.SL-e	WL	PP	fine	salt	calm/r	perp+par
	1981 research:						
18 '73	Vlissingenkan.	WL	PP/E	fine	fresh	calm	parallel
19 '73	Den Helder Sea	WL	PP	normal	salt	rough	perp+par

Nr. Year	Location	Site	Mat.	Subsoil	Water-quality	Hydr. circum	Current
20 '67	IJssel km 945	WL	PA	normal	fresh	rough	perp+par
21 '67	IJssel km 941	WL	PE	normal	fresh	rough	perp+par
22 '75	Friese seadike	Abv	PA	fine	salt	rough	perp+par
23 '75	Friese seadike	WL	PA	fine	salt	rough	perp+par
24 '70	Friese s,dike	WL	PP	fine	salt	rough	perp+par
25 '66	Baakse Beek	Abv	PA	fine	fresh	calm	perp+par
26 '72	Baakse Beek	WL	PE	fine	fresh	calm	perp+par
27 '66	Baakse Beek	WL	PA	fine	fresh	calm	perp+par
28 '69	Peelkan.v'cal	WL	PA	fine	fresh	calm	parallel
29 '69	Roggenplaat	WL	PP	fine	salt	rough	perp+par
30 '71	Maas km 230	Abv	PP	fine	fresh	calm/r	perp+par
31 '71	Maas km 230	WL	PP	fine	fresh	calm/r	perp+par
32 '72	R'dam port	Abv	PP	fine	salt	rough	perp+par
33 '72	R'dam port	WL	PP	fine	salt	rough	perp+par

Notes

1. Site; : WL = waterlevel, Abv = above waterlevel
2. Subsoil; : fine = D50<100 micron, normal = D50>100 micron.
3. Hydr. circum.: calm = Hs < 0.5 m + V < 2.5 m/sec. rough = Hs > 0.5 m or V > 2.5 m/sec.

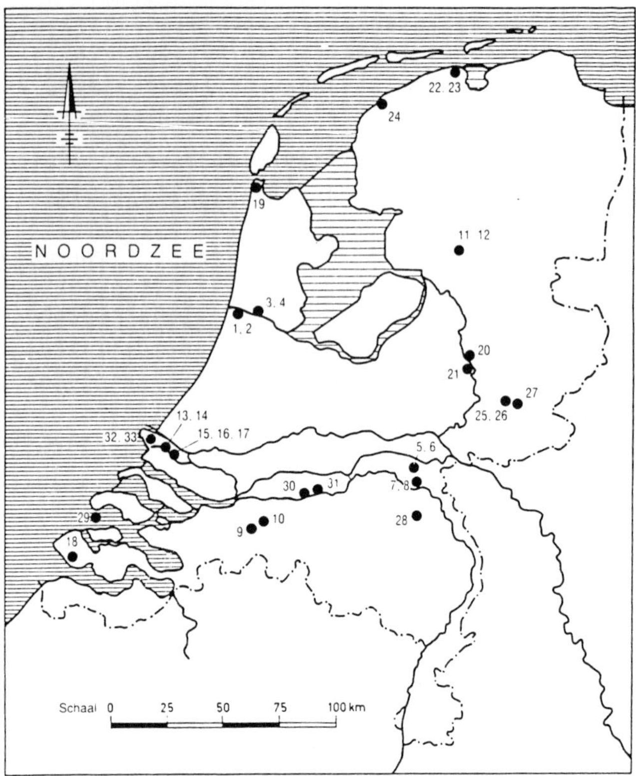

Fig. 1. Location plan

thoughts that it would be impossible to obtain undisturbed soil samples. Tidal conditions or sampling during low water periods made sampling possible. Finally it was found that at some of the 32 sites, what was thought to be a good location did not necessarily turn out to be so. Each location or even sampling site showed clearly its own character. Consequently the differences between sampling sites that were thought to be equal were large (See Results).

7. Realisation of sampling

There were 3 different phases

(a) preparation
(b) actual sampling
(c) repair.

7.1. Preparation

This involves taking away all material on top of the geotextile in such a way that there was no or minimal damaging of the geotextile and soil samples. Since there was a great variety of protection systems, the approach was not always the same. Riprap and slag were taken away largely by crane. Concrete or granite was removed by hand. Slag was shovelled manually into the crane bucket, as was the lowest layer of riprap (Fig. 2). Where there was a layer of gravel or tropical wood directly on the geotextile, this layer was kept in place, even during soil sampling. Only exactly on the spot where the tubes for the soil sampling were to be inserted into the slope was this layer manually removed. When there was no layer of gravel or wood, wooden planks were laid upon the base geotextile to prevent 'foot' damage. After sampling the soil the geotextile was fully uncovered, to take the geotextile samples. This work normally took a few hours, depending on the type of protection.

A number of different types of materials were found on the geotextile or

Fig. 2. Removal of armour stone

Fig. 3. 'De Stier' two tonne hydraulic jack used for taking undisturbed samples

the gravel, e.g. riprap on wood, concrete blocks, granite boulders, basalt, slag blocks or tiles, a bituminous consolidated natural stone (grauacke) and, in one case, a vertical wooden structure.

7.2. Sampling

Equipment available for this part of the work was: a 2-ton sounding equipment, hired from Soilmechanics Delft; a construction to anchor this equipment. This enabled the crew to move the equipment on rails 6 metres long, but to have it coupled and anchored sufficiently during installation of the tubes (Fig. 3).

The procedure was as follows.

Fig. 4. Insertion of sampling tool

Fig. 5. Thermal cutting

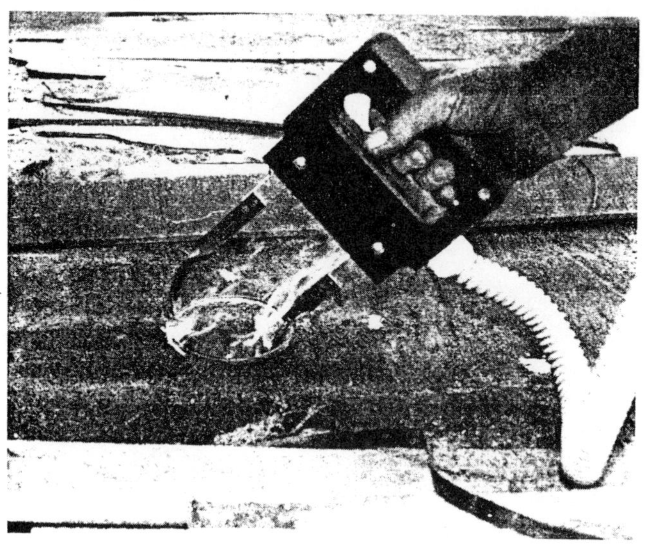

The soil samples were taken using perspex tubes, 100-110mm diameter, 200mm long, pressed into the soil and then pulled out. In this process the soil was undamaged. During/after the process the tubes were sealed to prevent damage (Fig. 4).

Prior to insertion, the geotextile was thermally cut to exactly the outside diameter of the soil collecting tubes (110mm) (Fig. 5).

Sometimes additional preparation was necessary when, due to wet conditions, the thermal cutting procedure failed.

After the geotextile had been cut the perspex tubes were pushed into the soil. 15 samples from each sampling site were taken in this manner. The tubes were filled above the sample with 20-30mm of polystyrene granules, then grease was placed around the edge for vacuum sealing.

With the same thermal cutting device larger geotextile samples were taken later in the process. The size of these pieces was about 5 square metres.

A unique coding system was devised for all the locations. Also at each location an undisturbed sample was taken, sealed into bags, packed into shockproof crates and sent to Delft Soilmechanics Laboratory. The geotextile sample was wrapped around a innercore, to prevent creasing, packed, sealed and shipped to TNO Delft (Fig. 6).

At 2 sites this procedure was not possible. On the Noordzeekanaal the geotextile was found on a 4:1 slope instead of the expected 1:3. The polypropylene fabric at that particular spot had not been permeable enough; the slope had bulged, but by dumping additional riprap the'problem' had been solved. Here the soil sampling had to be done by hand. The other exception was an inland canal where the vertical wooden structure was found.

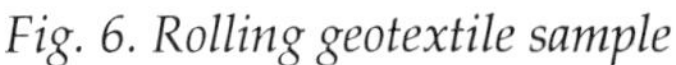

Fig. 6. Rolling geotextile sample

Fig. 7. Reinstatement of sampling area

During the work large numbers of photographs were taken to show under what conditions the samples were found and taken. The estimate of time for each sampling site turned out to be crucial, especially in tidal zones. But luckily it was always found to be correct in spite the large variety of conditions.

7.3. Repair

This part of the job started by refilling the slope after taking the samples. The replacement geotextile was then laid, and where fascines had originally been used they were replaced. Repairing hand-placed systems turned out to be most difficult. Repairs generally took half a day, but sometimes it took 1 - 1.5 days per site (Fig. 7).

All this work was done by the project leader and his assistant with the help of 2 men for preparation, sampling and repair. Occasionally a crane driver was needed and in all cases 1 man from Soilmechanics Delft was available to take the soil samples. Only at 2 sites was there a need for 2 extra men.

8. Laboratory research

In Section 2 mention was made of the purpose of the research. To obtain as much know-how as possible about ageing, clogging and blocking of geotextiles, it is necessary to set out a number of properties of the geotextile and the subsoil at the various sites.

(a) Ageing is a function of: the type of chemical, circumstances in which the geotextile had to perform, loads working on the geotextile, water quality, time.

(b) Clogging and blocking are a function of: type of geotextile (mesh), hydraulic conditions (waves, currents, groundwater, etc.), consistency of subsoil (particle diameter), water quality (also percentage of sludge), time.

To be able to determine the effect of these factors it is essential to know: the original properties of the geotextile, the present condition of the geotextile, what changes there were in the subsoil.

To determine the present state of the geotextile there were three possibilities

(a) fibre-technical research, focused on physical properties
(b) analytical-chemical research for the physical properties
(c) hydraulic laboratory research with regard to permeability and soil tightness of the geotextile.

The soil properties were established through

(a) soil mechanical research in 2 parts: analysis of some of the samples to find out the distribution of the particles at various depth under the filter; visual description of the samples to lay down the type, homogeneity, etc., of the soil.
(b) hydraulic research related to permeability of the soil whereupon the filter and the ballast were laid.

In this research programme the following Institutes participated: TNO Delft Fibre Institute, for the fibre technical research; TNO Delft Plastic and Rubber Institute for the chemical-analytical research; Delft Soil Mechanics Laboratory for soil mechanics; Delft Hydraulics (de Voorst) for the hydraulic research.

8.1. Fibre technical research.

Only when the original properties were known could the ageing could be determined. From a sample of 5-8 metre length, parallel to the river or canal, and 1.5 - 2 metre width perpendicular, 3 samples of 1 by 1 metre were taken by TNO Fibre Institute.

Physical properties were determined by the following investigations

(a) number of threads per 100mm in 2 directions
(b) mass divided by length, of 4 times 10 yarns, 1 metre long, taken from the 4 sides of the sample
(c) rate of loss of length of those yarns, due to the weaving process
(d) tensile strength and elongation at break. All these tests were done according to standardised (NEN) methods. It lead to many facts but in this

paper we will restrict ourselves to the conclusions.

8.2. *Analytical chemical research*

The purpose was to establish the chemical constitution of the material and to compare this with the original composition. This should give an indication of ageing and remaining life. The research was done by TNO Delft Plastic and Rubber Institute, from again 1 by 1 metre samples and involved

(a) infrared scanning
(b) establishing percentage of Polypropylene and/or Polyethylene
(c) establishing percentage of carbon black remaining in the yarn
(d) quantitative and qualitative analysis of anti-oxidants still present
(e) 'ovenlife' tests, for polypropylene and for polyethylene, and Additional Differential Scanning Calory tests, since the 'oven' life test proved not to be suitable.

Tests (b) and (c) were carried out twice.

In addition other tests were carried out on some geotextiles that were still available from old stock at the suppliers. This could only be done with a few geotextiles, due to unexpected changes which sometimes only came out during this testing.

8.3. *Soil mechanics research*

The purpose was to establish the composition of the subsoil directly under the geotextile. This supplies information on the behaviour of the geotextile as a filter; together with the hydraulic research it enables us to compare changes with changes in composition. It tells also whether the installed geotextile is still permeable and soiltight enough.

The 15 samples from each site were divided into 3 lots of 5. The 5 best properly filled tubes, i.e. no crevices along the wall or under the geotextile, were used for the hydraulic research. A second set was fully described in a grain/particle analysis in 4 layers, 0-5mm, 5-10mm, 10-15mm and 150-155mm under the geotextile. At each layer a particle size analysis was made.

The particle size analysis was ultimately done on the material used for the hydraulic research and on the remaining reserve samples.

8.4. *Hydraulic research*

This was done at Delft Hydraulics at de Voorst and it gave ideas about changes of permeability of the geotextile and the subsoil in relation to time. Also the soil tightness was established. 5 samples were used as stated earlier, and they were selected from the least disturbed. After testing soil and geotextile together the geotextile was tested separately. Tests were done in the original sampling tube and for that reason 6 holes were made in the tube at certain heights. This was to find the loss of pressure in the samples at those heights when testing. 3 tests were made with different pressure heads. The tests were concluded when a stationary situation was reached.

D'Arcy's law was applied to find the k-value. Again in the context of this report only the conclusions are stated but the testing was carried out fully in accordance with standardised methods.

8.5. Interpretation of results

There are a number of factors for judging the behaviour of geotextile coming from both soil mechanic and hydraulic research. Please note that the geotextile was not only tested as found with and without soil underneath, but also cleaned and in some cases as new.

Conclusions were drawn on the basis of 5 factors.

The shape of the pressure line of the soil sample from the hydraulic tests. These were either convex, meaning the permeability decreases going from bottom to top when a relatively tight ballast is applied, straight when conditions are calm and balanced indicating no change in permeability and finally concave, when the permeability increases closer to the geotextile due to its open structure, not silted up and with an open top layer.

Factor F1, indicating the permeability of the top 5mm soil layer.

Factor F2,indicating the rate of 'pollution' of the geotextile.

Factor F3, indicating the possibility of overpressure.

Factor F4, indicating where exactly this overpressure will take place, either directly underneath the geotextile or under the top 5mm soil layer.

In addition all sorts of cross calculations were done to be more sure of the results and consequently of the conclusions. This report is intended to show the results of the tests and we apologise for the rather brief descriptions above of the test procedures.

9. Locations of sampling

We will not, for brevity's sake, include this section of the original report. Suffice to say that for meaningful results the locations of the sampling must be chosen carefully.

10. Results per laboratory

As above only the conclusions of the report are given here.

10.1. *Fibre technical research*

The geotextile can differ in properties per location because of

(a) different qualities can be found on the same site
(b) more than one type of yarn is used in a geotextile
(c) larger than expected spread of results of property tests of the yarns
(d) local influences like load, crimp, dirt and damage.

The relative specific strength, defined as the quotient of specific tensile strength of samples taken and original material is for the various raw materials

polypropylene 0.6 - 0.9
polyethylene 0.7 - 0.9
polyamide 0.5 - 0.7

The same quotient for elongation could not be given due to the wide spread of results.

Decrease of tensile strength is basically the result of mechanical damage, almost always caused by transport and handling on the site during application.

The position of the geotextiles above or on the waterline: different hydraulic circumstances and even exposure time did not give significant differences.

10.2. *Analytical chemical research*

The geotextiles can differ in properties per location due to

(a) different qualities on one site
(b) more than one type of yarn in one quality of geotextile
(c) larger than expected spread in properties of the yarn.

The remaining percentage of anti-oxidants in the polypropylene is low when some is left in the first place. Polyethylene and polyamide were not treated with anti-oxidants originally.

No difference was found in loss of anti-oxidants for samples of the two different exposure sites, above or at the water level.

Soils containing iron shorten the lifetime of polypropylene considerably. Acid conditions work negatively on anti-oxidants.

Taking in account the eventual existence of iron-containing soils, the lifetime expectancy of polypropylene under thermo-oxydative conditions is about 15 - 30 years. Polyethylene has a lifetime expectancy that is at least 20 times longer under thermo-oxydative conditions. For polyamide thermo-oxydation has no effect.

10.3. Soil mechanical research

At each of the sites the 15 samples taken, were visually described. On 5 of them a particle size analysis was done in 4 separate layers of 5mm thickness in a set pattern.

The descriptions give information about the homogeneity of the samples, the existence of silted layers, the type of soil, etc. This information is of importance to interpret the hydraulic research. This is also the case for the particle analysis.

Silted layers were found in samples, taken from underneath geotextiles with the largest mesh (mesh fabric).

In some locations where outflow of groundwater takes place, a natural filter was found. The other conditions like subsoil, characteristics of mesh etc. were different.

Loss of subsoil particles endangering the stability of the construction were not found.

10.4. Hydraulic research

Five factors are used to establish the results as described above. The conclusions drawn from all this areas follows.

(a) The influence of the 5 mm layer directly underneath the geotextile is crucial for the functioning of the total structure. This layer can be less permeable than the original, especially in case of a silted top layer.

(b) Siltation is the result of lack of 'wash'. For that reason it is bigger at the point above the water level than right on it or below.

(c) Potential danger for overpressure comes from the 5 mm layer when it is silted. The influence of clogged geotextile (all wovens in this research) is far less than the influence of this particular layer. Only in the case of soils containing iron will the geotextile lose permeability to such a degree that it equals the effect of the top layer of soil.

(d) The stability of the construction was not endangered by overpressure in any case in this research.

(e) Although in a large number of cases the soil tightness was not according the standard design rules (Ogink, WL Method), no disadvantages were found. It proves that those rules are for the extremes.

(f) Almost all geotextiles are vulnerable to blocking but once again no disadvantageous effects were found.

(g) When designing the filter construction is is important that discharge of water and (eventually) the finest soil particles can take place. The geotextile should be quite open. 090 / D90 between 1 and 2 is possible for the 32 Dutch sites at least.

11. References

OGNIK, H. J. M. *Investigations on the Hydraulic Characteristics of Synthetic Fabrics*, Publication 146 Delft, Delft Hydraulics Laboratory, 1975.

Kuunststoffilters in Kust & en Oeverwerken, Rotterdam, Nederlanse Vereniging Kust-En Oeverwerken, 1989.

Experimental evaluation of filter performance for geotextiles in landfills

D. CAZZUFFI, ENEL Research Centre for Hydraulics and Structures, Milan, Italy, and R. COSSU, Institute of Hydraulics, University of Cagliari

Geotextiles and geocomposites are often used in landfills with filter and drainage functions. The paper reports the results of permeability tests, using leachate, carried out on several kinds of synthetic materials. Mechanical and biological clogging have been investigated.

1. Introduction

A filter in a drainage system represents the key factor in landfill design, management and environmental impact. The purpose of an efficient drainage system is to allow a quick flow and collection of leachate, avoiding formation of an hydraulic head above the landfill bottom and of perched water tables. When these phenomena occur, the following risks arise

(a) leachate diffusion along the slopes or through eventual lining failures
(b) difficulty in biogas abstraction
(c) stability problems, particularly when the landfilling is being carried out by monding or on a slope.

These remarks imply that materials for filters in landfill drainage systems should be carefully considered and studied: in this field, in fact, the utilisation of geotextiles and geocomposites as filters has very often been handled without proper consideration of the performance of these synthetic products in landfill environment. The clogging of geotextiles due to leachate has been already investigated both for mechanical and biological effects (Cazzuffi and Cancelli, 1987; Koerner and Koerner, 1989). In this regard, specific design criteria have been proposed (Rollin and Denis, 1987).

The research presented here was carried out in order to further investigate the hydraulic behaviour to leachate of a wide range of products. Several

Table 1. Testing programme

RUN	TESTED GEOTEXTILES	TEST METHODOLOGY
I	nonwoven geotextiles	short-term permittivity test
II	woven geotextiles and geocomposites	long-term permittivity test
III	nonwoven and woven geotextiles; geocomposites	short-term permittivity test after long-term leachate exposition: a = 2 month b = 3 month c = 5 month

geotextiles and filtering geocomposites, which are representative of the most widely used types of synthetic materials used for landfill drainage systems, are considered.

2. Methodology

The research included carrying out permittivity tests on different kinds of geotextiles and geocomposites commercially available in Italy. Three experimental runs were planned using the programme shown in Table 1.

2.1. Testing apparatus

All the tests were carried out using a permeameter which had been specifically designed for geotextiles and constructed at the ENEL-CRIS Special Materials laboratory in Milan. Most parts are made of plexiglass: allowing direct observation of all phenomena occurring during permeability tests. The hydraulic layout of apparatus is shown in Fig. 1; the basic elements are

(a) the permeameter
(b) the feed section
(c) the flow measuring section.

The permeameter is formed of 3 coaxial cylinders with internal diameters of

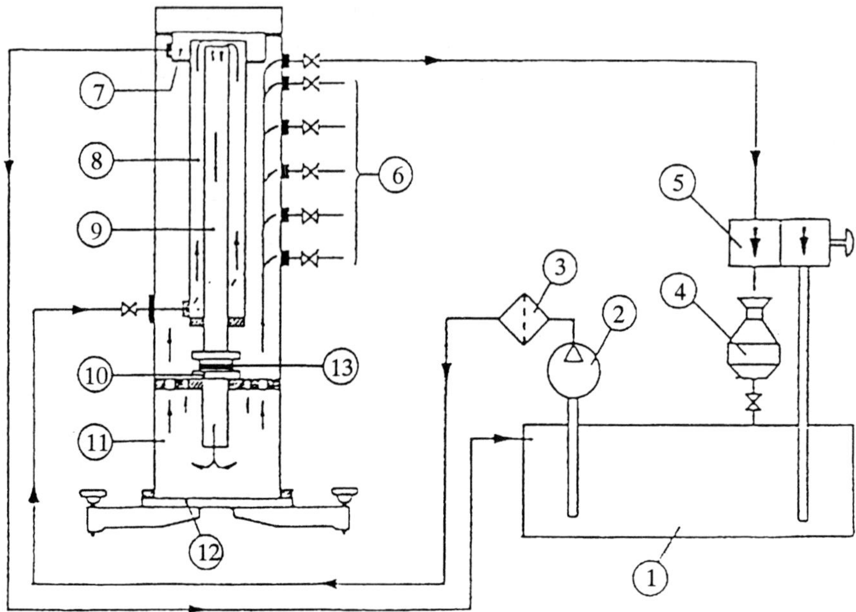

Fig. 1. Hydraulic scheme of the closed circuit, constant head permeameter for permittivity tests: (1) liquid tank, (2) feed pump, (3) filter, (4) calibrated vessel, (5) two way valve, (6) outlets, (7) overflow system, (8) constant level cylinder, (9) feed cylinder, (10) specimen bearing, (11) external cylinder, (12) base, 913) geotextile specimen

284, 124 and 51 mm respectively. The inner cylinder houses, at its base, the geotextile specimen to be tested and simultaneously acts as feeding cylinder for the permeability test. The intermediate cylinder acts as a constant level cylinder; the water flowing from the spillway is directed, by means of an overflow pipe, to the feeding tank. The external cylinder provides support for the apparatus; 6 different values of hydrostatic load can be applied (h is given by the difference between the spillway level and the selected outlet and is maintained constant during a single test). Each opening has a cross section of 100 mm by 12 mm. The feeding apparatus consists of a pump which sucks water from a feed tank; this tank also represents the backwater tank at the end of the test circuit. The flow measuring section is formed by a calibrated reservoir that can be fed or excluded by a two-way valve.

This constant head permeameter has a closed hydraulic circuit so that the same permeability test can be performed by using the same small quantity of liquid (demineralised de-aired water or leachate).

When performing normal permeability tests on geotextiles with water, a particular problem is caused by the validity limits of Darcy-Ritter's law

(Francia 1983; Gourc *et al.* 1982). For this reason a minimum value of the specimen thickness H, in turn a function of the hydrostatic load h, should be maintained. When testing materials with water, the minimum, value of H was obtained by superimposing different layers of the same geotextile; the number of layers ranged from 3 to 6, depending on the thickness T_g of the geotextile or of the geocomposite to be tested. When using leachate, laminar flow conditions were observed and only one layer of material was used.

The coefficient of normal permeability k_n and the permittivity Ψ of the geotextile are computed by the formulas

$$k_n = VH/Aht = VxT_g/Aht \quad (1)$$

$$\Psi = Vx/Aht \quad (2)$$

where V = calibrated volume = 1450×10^{-6} m^3, H = specimen thickness (m), x = number of geotextile layers forming a specimen, h = hydrostatic load (m), t = time required to fill the calibrated volume V during a single test (s), T_g= geotextile or geocomposite thickness (m) and A = specimen surface = 2043×10^{-6} m^2.

The testing apparatus, with simple modifications, also allows testing under a normal compressive stress. However, in this study only tests with no normal stress have been performed in order to investigate the least critical conditions.

2.2. Materials

The materials used in the experiments included geotextiles, geocomposites, leachate and demineralised de-aired water.

Table 2. Analytical compositions of the different leachates used in different runs of the experimental tests (see Table 1)

PARAMETER	L1 (I)	L2 (II)	L3 (IIIa)	L4 (IIIb)	L5 (IIIc)	L6 (imm)
pH	6/6.3	7.2/7.3	7.1/7.4	8.1/8.9	7.7/8.4	7.1/8.1
COD (g/l)	28.8/38.5	13.6/35.4	31.9/34.4	7.3/11.7	19.1/20.3	8.8/32.0
BOD (g/l)	3.0/10.4	5.0/11.0	13.0/14.5	2.0/3.7	6.4/7.8	3.4/13.0
TSS	12.4/20.5	11.6/21.3	20.2/22.7	12.1/12.4	16.2/16.5	9.1/22.7
TVS	5.0/9.7	4.6/10.6	10.2/11.7	4.9/5.4	7.6/7.9	3.8/11.7

Table 3. General characteristic of materials tested in the different runs (see Table 1)

RUN	REFERENCE	NAME	TYPE	MATERIAL	GEOTEXTILE PROCESS	μ (g/m^2)	T_g (mm)	k_{w20}
I	A	Bidim U34	nonwoven	PET	NP,CF	290	2.9	$4.1.10^{-3}$
I	B	Drefon S45	nonwoven	PET	NP,SF	200	2.2	$4.5.10^{-3}$
I	C	Tecnofelt FAG	nonwoven	PET	NP,SF	300	3.4	$5.9.10^{-3}$
I	E	Polyfelt TS 750	nonwoven	PP	NP,CF	370	3.0	$2.8.10^{-3}$
I	F	Drefon SIA 200	nonwoven	PP	NP,SF	200	2.9	$5.0.10^{-3}$
I III	G	Terram 1000	nonwoven	PP PE	TB,CF	140	0.8	$1.7.10^{-3}$
I	H	Typar 3807	nonwoven	PP	TB,CF	280	0.7	$1.7.10^{-4}$
I	I	Drefon SIA 400	nonwoven	PP	NP,SF	400	5.2	$5.0.10^{-3}$
I	L	Stratum	nonwoven	PP	NP,SF	450	5.4	$1.2.10^{-3}$
II	M	Terram W/3-3	woven	PP	W	206	0.8	$3.0.10^{-2}$
II	N	Terram W/7-7	woven	PP	W	485	1.4	$2.5.10^{-3}$
II III	O	Terram W/12-12	woven	PP	W	794	2.1	$1.1.10^{-3}$
II	R	Tecnodrain	geocomposite (nonwoven+geonet)	PP-PVC	TB,CF	863	6.4	$5.0.10^{-3}$
II	S	Tigerdrain	geocomposite (nonwoven+geonet)	PP-PE	NP,SF	1177	17.5	$7.3.10^{-3}$
II	T	Filtram 1BZ	geocomposite (nonwoven+geonet) + nonwoven) geocomposite	PP-PE	TB,CF	1256	6.1	$5.4.10^{-4}$

II	U	Filtram 1BZ	(nonwoven+geonet) + nonwoven)	PP-PE	TB,CF	1321	12.5	$3.6.10^{-4}$
II	V	Terram 7M7/700	geocomposite (nonwoven+geonet) + nonwoven)	PP-PE	TB,CF	739	2.2	$4.0.10^{-4}$
III	Z	Terram W/20-4	woven	PP	W	594	1.8	$2.2.10^{-4}$
III	W	Polyfelt TS 700	nonwoven	PP	NP,CF	306	2.4	$7.1.10^{-3}$
III	Y	Terbond A-300	nonwoven	PET	NP,SF	419	3.8	$3.6.10^{-2}$

μ = mass per unit area; T_g = thickness; k_{w20} = permeability to water at standard temperature (20°C).
PET = polyester; PP = polypropylene; PE = polyethilene; PVC = polivynilcloride; NP = needle-punched; TB = thermo-bonded; CF = continuous filament; SF = staple-fibre; W = woven.

Leachates. The leachate used for the tests was collected at the sanitary landfill of Mariano Comense (Como). The tests were carried out in different periods and the leachate composition reflects typical variation for this waste-water.

In Table 2 the range of the analytical composition of leachates used in the different experimental runs is reported. Leachate L1 is a typical acidic phase leachate, while the others are typical of an unstable methanogenic phase. Leachate L6 was used for the long-term exposure to leachate of geotextiles before conducting the permittivity test, in research phase III.

Geotextiles and geocomposites. The complete list of geotextiles and geo-composites tested is reported in Table 3, together with the most significant physical properties. Mass per unit area (μ) and thickness (T_g) were measured in standard conditions. Permeability to water (k_{w20}) was also measured at standard temperature of 20°C. The values reported in Table 3 are the averages of 7-10 measurements.

The list includes nonwoven and woven geotextiles. The nonwovens were mostly needled-punched and thermobonded. Among needled-punched ge-otextiles, some staple fibre Italian products were considered. The mass per unit area ranged from 140 to 800 g/m^2 and the nominal thickness from 0.7 to 5.4 mm for non-composite geotextiles.

The list also includes geocomposites made by bonding nonwoven geotex-tiles to geonets (and in one case to woven geotextiles). The mass per unit area and the nominal thickness of these materials are also reported in Table 3.

2.3. Experiment procedure

Test runs I and II. The geotextile specimens were tested using the apparatus previously described. Leachate was circulated through the permeameter until stable conditions were achieved. During runs I, when nonwoven geotextiles were studied, the test durations were in the range 40 to 300 min.

Permeability measurements during the tests were carried out periodically, the frequency dependent upon the rate of clogging of the individual geotextile. A higher frequency of measurements was adopted for the less permeable geotextiles (Pessina, 1988).

After each test the leachate was changed and a fresh leachate was used for the next test.

Test runs III. In the third runs specimens of geotextiles and geocomposites were immersed in leachate, in anaerobic conditions, to look at the effects of biological growth on clogging.

The specimens were immersed in a stainless steel vessel with a volume of 50 l, hung from the top of the vessel. The vessel was thermo-insulated at 30°C. Three immersion periods were used

(a) 2 months
(b) 3 months
(c) 5 months.

After immersion the specimens were subjected to a short-term permittivity test using leachate in the permeameter previously described. The same test was carried out on an unused clean specimen with the same characteristics .

The same duration (300 min) was used for all of these tests, not the variable durations used in the test runs I and II.

After leachate circulation the specimens were analysed so that the amount of trapped total and volatile solids could be measured (Ferruti, 1991).

Additional rectangular specimens (300 mm × 170 mm in size) of the different materials were immersed in the leachate immersion cell. These specimens were used for mechanical testing.

Mechanical testing to establish the tensile strength and strain at failure were performed on two test frames. The two test frames had different maximum load capacities: JJ Instruments, mod. M5K (max. load 5 kN) and Galdabini mod. PMA/20 (max. load 200 kN).

3. Results and discussion

The results of permeability tests on nonwoven geotextiles are summarised in Fig. 2, a plot of k against time. All test materials show a marked decrease of the coefficient of permeability k with elapsed time.

The reduction from the initial to final permeability lies between 100 and

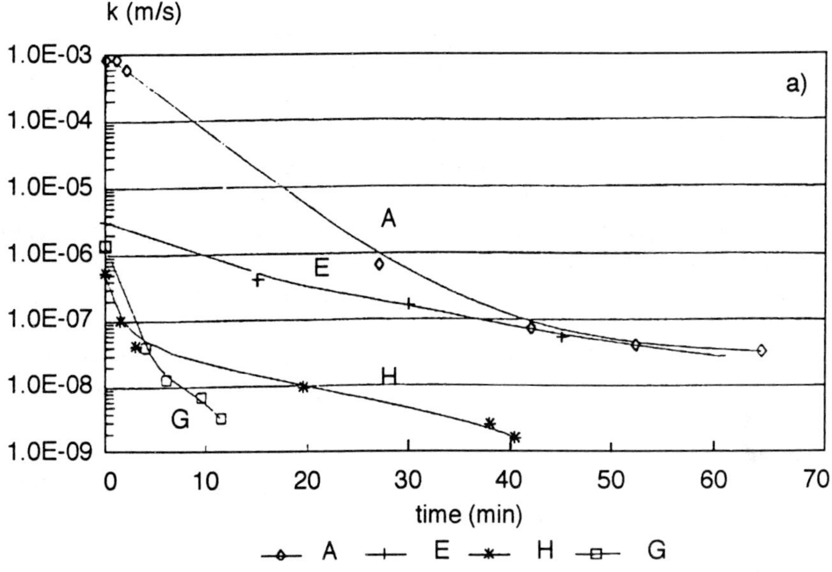

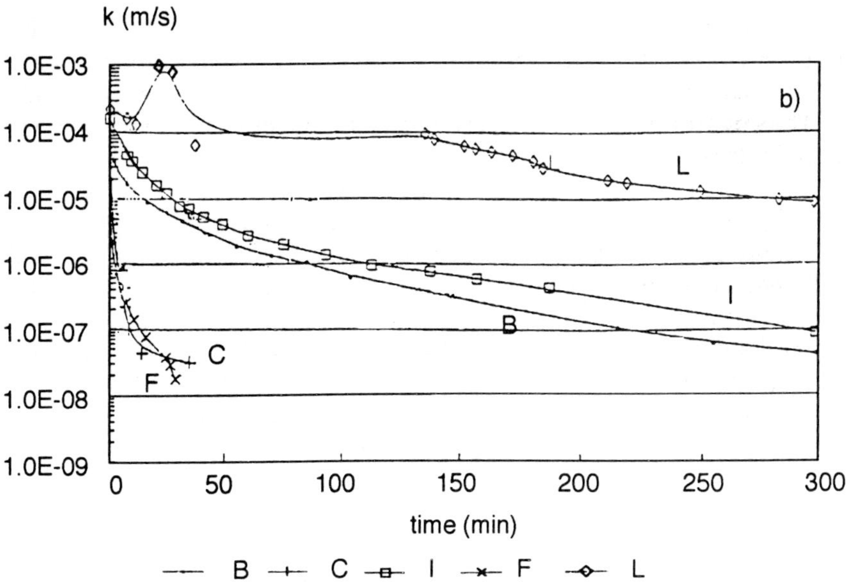

Fig. 2. Permeability to leachate against time observed in the permittivity tests on nonwoven geotextiles (I experimental run): (a) continuous filament, (b) staple fibre

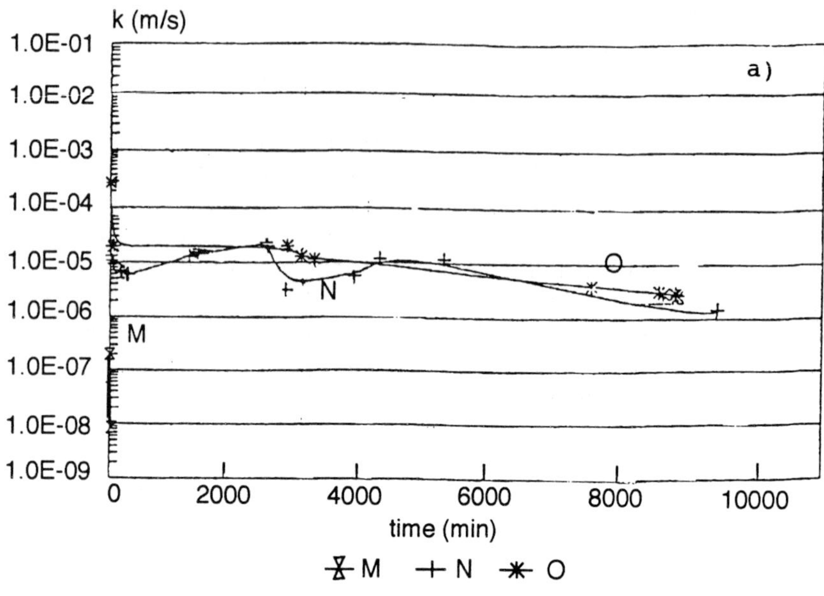

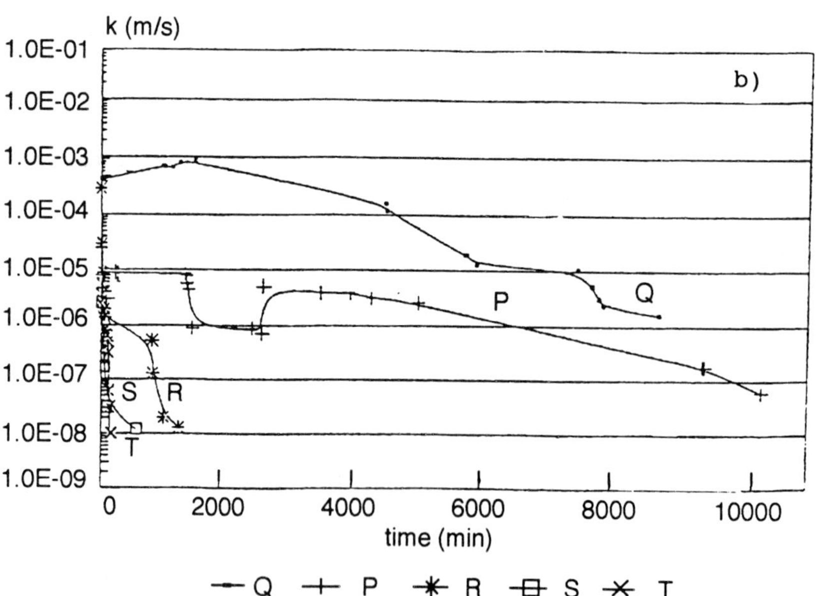

Fig. 3. Permeability to leachate against time observed in the permittivity tests in the II experimental run: (a) woven geotextiles, (b) geocomposites

1000 times. Comparing the values of permeability to leachate at the end of the test with the permeability to water for the different geotextiles, it can be seen that the effect of leachate is to reduce normal permeability to less than 1/100 000 of the original value.

The results obtained in the experimental run II using woven geotextiles and geocomposites are reported in Fig. 3. As already emphasised by Cazzuffi *et al.* (1991), generally woven geotextiles showed a good resistance to clogging, higher than that of nonwovens. Only one woven geotextile (reference M), with a close-weave structure, showed very rapid and intense clogging.

The composite geotextiles, including thermo-bonded nonwoven geotextiles as filters (references R, S and T), showed quicker clogging compared with geocomposites formed by a sandwich of needle-punched nonwoven geotextiles and plastic core (references P and Q).

In the experimental run III all the tests were performed with a duration of 300 min, as previously described.

The graphs in Fig. 4 refer to different immersion times in the leachate (T2, T3, T5) and to the corresponding non-immersed geotextile specimens.

From these results it is possible to observe a permeability decrease of all geotextiles when compared with their permeability to water (Fig. 5). As in the test run II, the wovens (sample references 0 and Z) showed a better resistance to clogging. The permeability decrease is influenced both by the solids content in the tested leachate and by the immersion time in the leachate (biological clogging).

Generally the permeability to leachate of the immersed samples was lower than the permeability of the non-immersed sample, with some exceptions, mainly the woven geotextiles (sample references O and Z), Fig. 5.

During test runs III the greatest clogging effect was observed for some nonwoven geotextiles (references U and V).

A linear relationship between decrease of permeability and immersion time in leachate cannot be clearly observed. The solids content in the leachates used varied, and was particularly was high for the run after two months immersion (Table 3).

The change in leachate properties is a general problem when performing long-term tests. The leachate changes dramatically when stored at low temperature or during percolation from the same landfill. The graphs in Fig. 6 clearly show how the specific permeability (ratio between permeability to leachate and permeability to water) depends on the solids content of the leachate and how this effect is increased by biological growth which occurs during immersion in leachate.

The quantity of total and volatile suspended solids trapped in the tested specimens is reported in Fig. 7.

Figure 8 shows the ratio between the solids trapped in the immersed geotextiles and in the non-immersed one, after the permittivity tests with

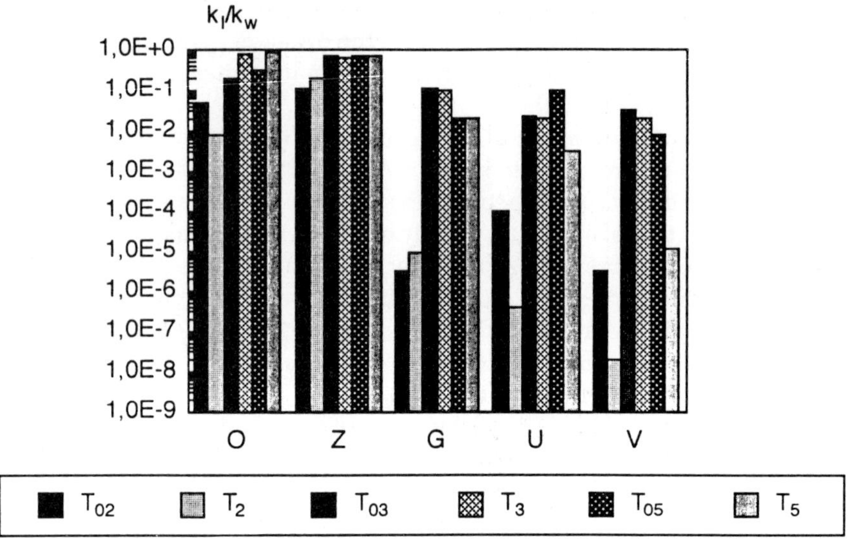

Fig. 4. Calculated values of the permeability ratio (k_l/k_w) for the different geotextiles used in the III experimental run. k_l = permeability to leachate, k_w = permeability to water on a virgin sample, T_{oi} = non-immersed sample tested after i months, T_i = immersed sample tested after i months

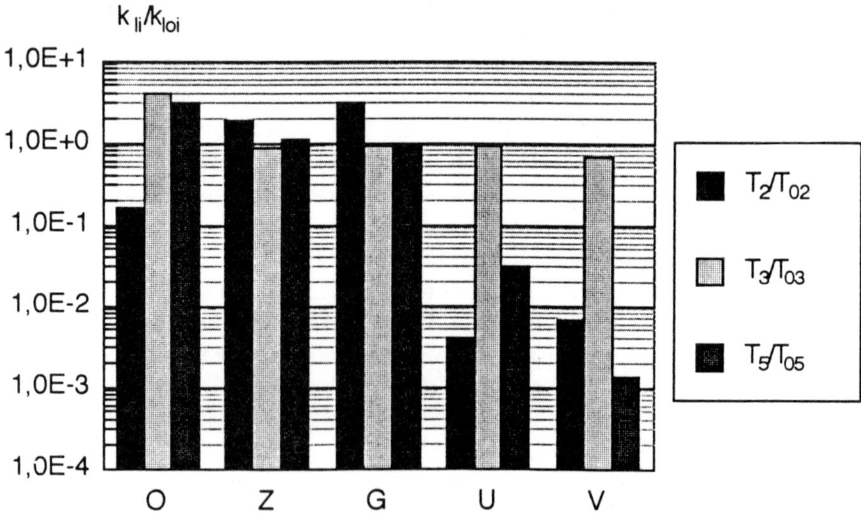

Fig. 5. Ratio between permeability to leachate of immersed (k_{li}) and non-immersed samples (k_{loi}) for the different (i) geotextiles tested in the III experimental run

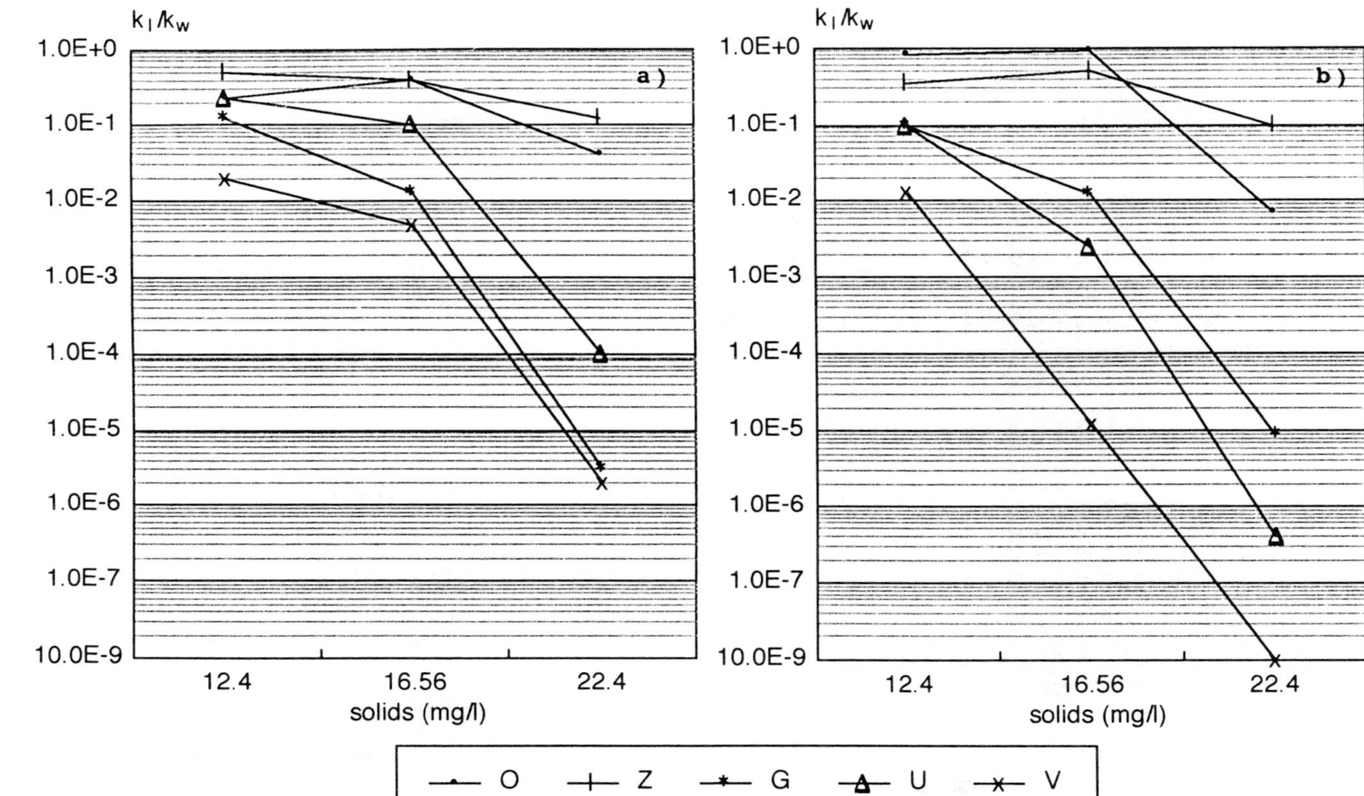

Fig. 6. Permeability ratio (k_p/k_w) of (a) non-immersed and (b) immersed specimens leachates used in the III experimental run against total solids content in the testing

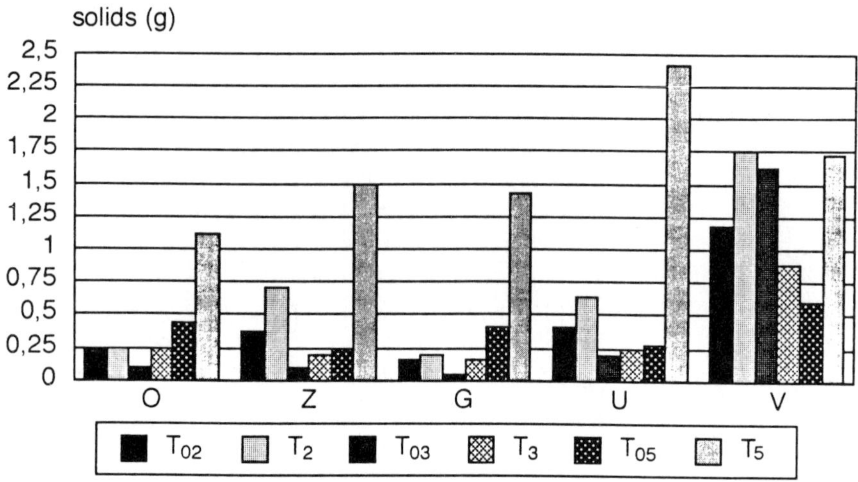

Fig. 7. Total and volatile solids trapped on the specimens of the different geotextiles used in the III experimental run. T_{oi} *= non-immersed sample tested after i months,* T_i *= immersed sample tested after i months*

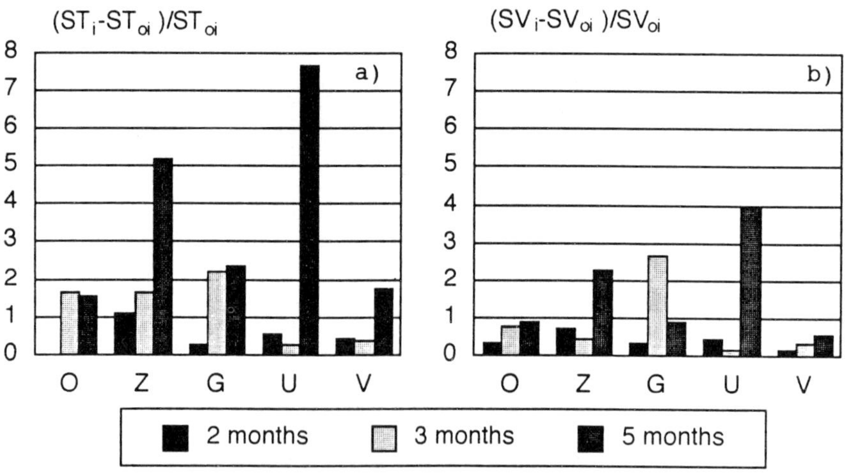

Fig. 8. Accumulation of (a) total and (b) volatile solids in the geotextile specimens, after different immersion times. ST_{oi} *and* SV_{oi} *= total and volatile solids trapped in the non-immersed specimens after the permittivity test with leachate.* ST_i *and* SV_i *= total and volatile solids trapped in the immersed specimens after the permittivity test with leachate*

a) Woven geotextiles: Terram W/12-12 (ref.O) and Terram W/20-4 (ref.Z).

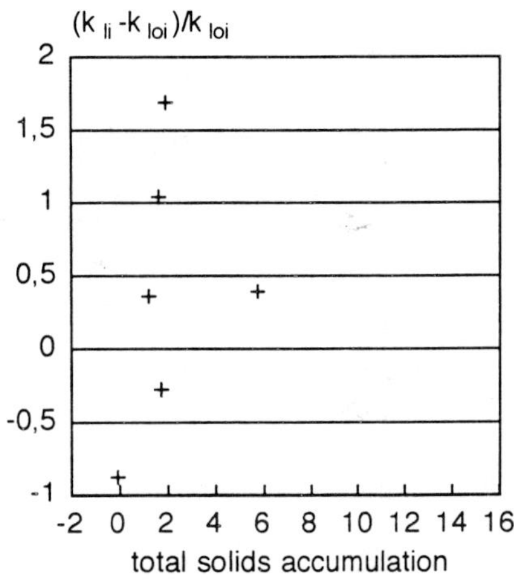

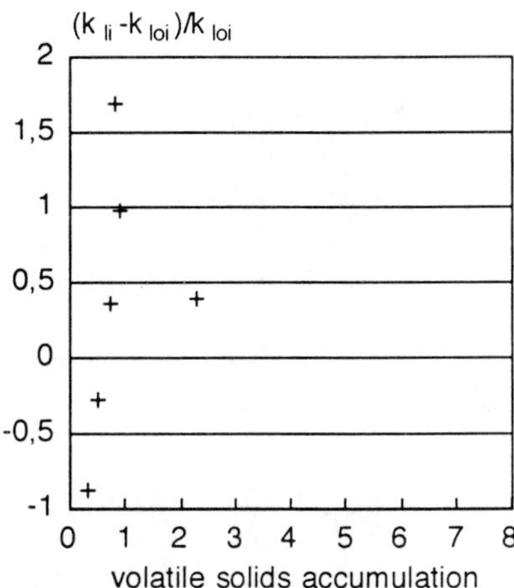

b) Nonwoven geotextiles: Terram 1000 (ref. G), Polyfelt TS700 (ref. U) and Terbond A 300 (ref. V).

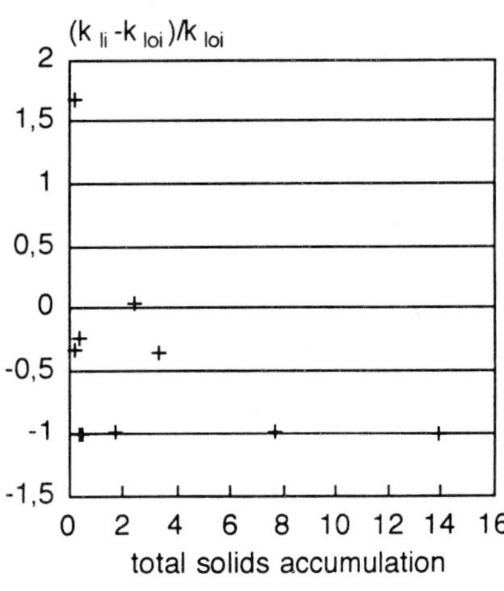

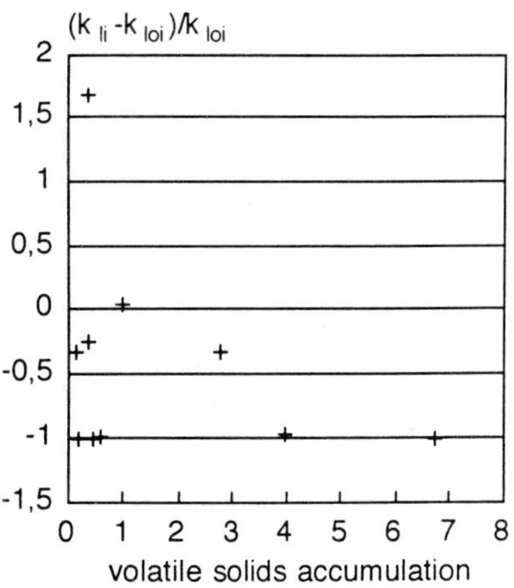

Fig. 9. Specific permeability variation against total and volatile solids accumulation (as defined in Fig. 8 caption), observed in the III experimental run, for woven and nonwoven geotextiles. k_{loi} *= permeability to leachate of non-immersed specimens.* k_{li} *= permeability to leachate of immersed specimens*

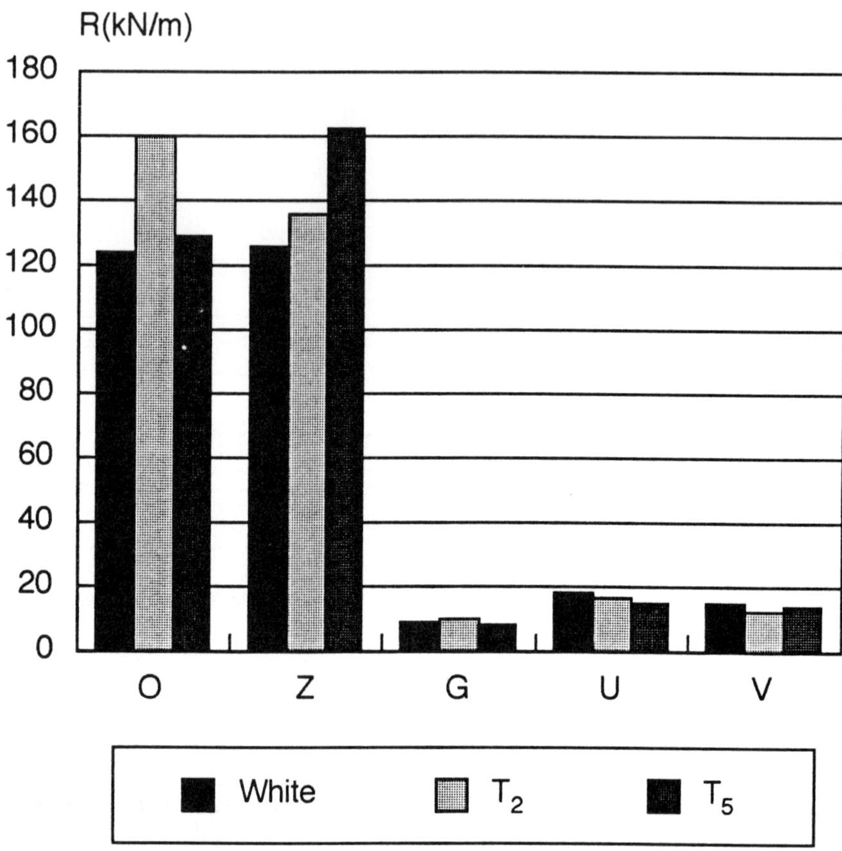

Fig. 10. Tensile strength (R) values measured in tensile tests on different geotextiles, after 2 and 5 months (T2, T5) of leachate immersion, and on virgin specimens of the same materials

leachate.

Coupling this observation with the results of the cycles I and II it is possible to hypothesize that there is a surface and a volumetric clogging. The geotextiles with higher specific volume generally have a higher solid accumulation capacity and consequently higher resistance to clogging phenomena. From another point of view the open structure of some geotextiles (as in some wovens) shows lower accumulations but also lower filtering effects, so that their resistance to clogging is higher. The effect can be seen in Fig. 9, where the variation of permeability to leachate of immersed and non-immersed geotextiles is related to the accumulation of total and volatile solids in the specimen. In woven geotextiles, permeability variation appears not to be related to the solids accumulation following immersion in leachate (Fig. 9).

On the contrary, the correlation is evident for the nonwoven geotextiles (Fig. 9).

Figures 10 and 11 show the results of the mechanical tests on the immersed and non-immersed geotextiles, tensile strength and strain at failure. The effect of leachate immersion on the tensile strength and strain at failure is small and insignificant.

4. Conclusions

The results of three cycles of tests on permeability to leachate of different kinds of geotextiles and geocomposites lead to the following conclusions.

Sanitary landfill leachate causes a marked decrease of permeability in

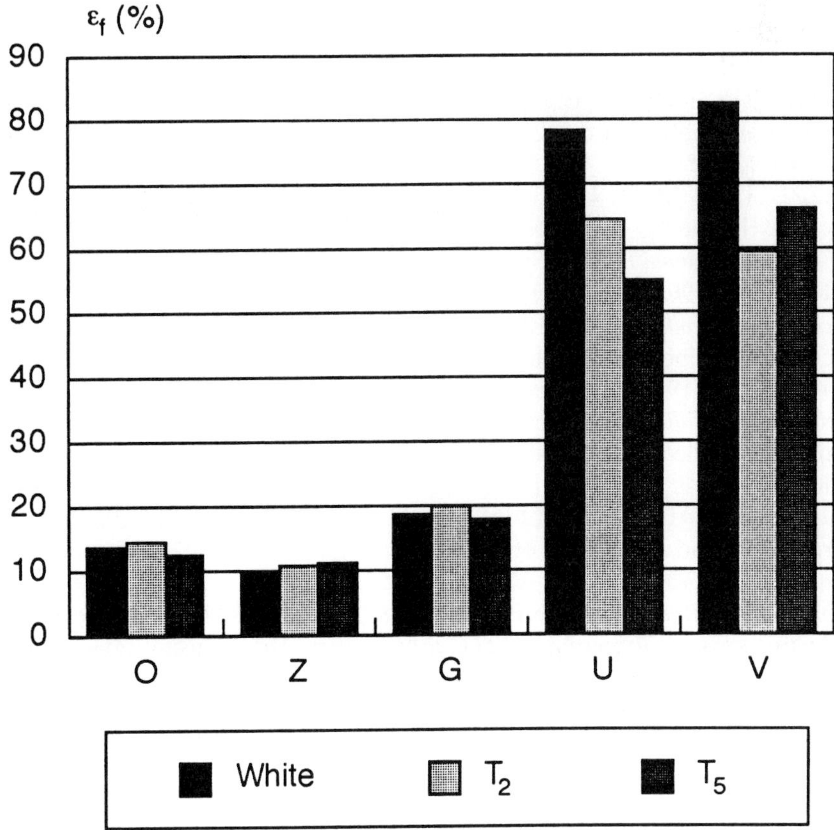

Fig. 11. Strain at failure (ε_f) values measured in tensile tests on different geotextiles, after 2 and 5 months (T2, T5) of leachate immersion, and on virgin specimens of the same materials

every kind of geotextile, under combined effects of surface deposition and volumetric accumulation of solids, both those contained in the leachate itself and those formed by biological growth.

The performance of geocomposites is dependent on the behaviour of the geotextile filter component with leachate. For instance, a sandwich including thermo-bonded nonwoven geotextiles shows the same problems of clogging as observed for the source products.

The mechanical strengths of geotextiles (tensile strength and strain at failure) are not significantly affected by exposure to leachate.

According to Rowe (1992) it is possible to conclude that geotextiles and geocomposites can perform a useful role as filters in landfill leachate collection systems, provided that in their design the potential for clogging is adequately recognised and allowed for.

5. References

CAZZUFI D. and CANCELLI A. (1987) Permittivity of geotextiles in presence of water and pollutant fluids. *Proceedings Geosynthetics 187 Conference*, New Orleans, 471-481.

CAZZUFI D., COSSU R., FERRUTI L. and LARANGDO C. (1991) Efficiency of geotextiles and geocomposites in landfill drainage systems. *Proceedings of the 3rd International Landfill Symposium*. Sardinia, vol. 1, 759-780.

FERRUTI L. (1991) *Indagine sperimentale sul comportamento dei geotessili nelle discariche controllate di RSU in presenza di attivita biologica*. Master Thesis, Faculty of Engineering, Politecnico di Milano.

FRANCIA L. (1983) *Studi sperimentali sui geotessili per applicazioni in campo geotecnico ed idraulico* (Experimental studies on geotextiles in geotechnical and hydraulic applications). Master Thesis, Faculty of Engineering, University of Bologna.

GOURC J.P., FAURE Y., ROLLIN A. and LAFLEUR J. (1982) Standard test of permittivity and application of Darcy's formula. *Proceedings 2nd International Conference on Geotextiles*, Las Vegas, 149-154.

KOERNER G.R. and KOERNER R.M. (1988) Biological clogging in leachate collection system. *Proceedings 2nd G.R.I. Seminar Durability and aging of geosynthetics*, Drexel University, Philadelphia, 1-18.

PESSINA D. (1988) *Indagine sperimentale sul comportamento di geotessili in impianti di scarico controllato di rifiuti solidi urbani*. Master Thesis, Faculty of Engineering, Politecnico di Milano.

ROLLIN A. and DENIS R. (1987) Geosynthetic filtration in landfill design. *Proceedings Geosynthetic 87 Conference*, New Orleans, 456-470.

ROWE K. (1992) Geotextiles in filtration and drainage: some challenging applications. *Geotextiles in filtration and drainage*, Thomas Telford, London, 1993.

Geotextile durability: current situation regarding test methods and standards

J. H. GREENWOOD, ERA Technology Ltd, Leatherhead, UK, and K. C. BRADYand G. R. A. WATTS, Transport Research Laboratory, Crowthorne, UK

1. Introduction

Geotextiles are being increasingly used in Civil Engineering works and, in common with the introduction of other novel materials and techniques, their usage has generally preceded the establishment of standardized test methods and specifications. The demand for the necessary documentation has varied according to the popularity and importance of the application but, understandably, has been more insistent with longer term structural applications such as reinforced soils. The design life currently specified by the Department of Transport for reinforced soil structures is 120 years, and so the durability of geotextile reinforcements is a major consideration for a designer.

The lack of long term performance data for a new material or technique can prevent its widespread adoption, but more usually promotes the use of accelerated testing methods. However without field verification, it is extremely difficult to translate data from short term tests into standards and specifications for long term usage.

The durability of a geotextile could be assessed from the results of a series of tests, each test aimed at defining the resistance to a particular form of degradation. To appreciate the significance of the results of such tests requires an understanding of the possible modes of failure of the geotextile system, and also which of the environmental conditions are critical to the particular application. Moreover some aspects of durability, such as the clogging of filters by biological agents, can only be properly investigated by field trials.

This paper provides descriptions of the various mechanisms of degradation for the geotextiles, and the methods used to characterize the degree of degradation. It also provides details of recent developments in the testing of geotextiles, including the publication of Standards and Codes of Practice.

Geotextiles in filtration and drainage Thomas Telford, London, 1993

2. Types of geotextile

In this paper, the term 'geotextile' is used to describe a range of products that include: woven and non-woven fabrics, extruded meshes and grids, and strip elements: it excludes geomembranes as these are essentially impermeable to the passage of water.

Geotextiles have been most commonly manufactured from polyethylene, polypropylene, polyester and polyamide, but other polymers such as polyvinyl chloride have also been used as coatings. The most common types are chosen because of their availability, economy, and the potential for engineering their properties by chemical and physical means to serve a particular purpose.

Geotextiles can be manufactured using a variety of techniques. Fabrics are essentially formed in a two stage process: the first concerns the manufacture of the linear components, such as filaments and tapes, and the second the assembly of these components using processes such as knitting, weaving, needle-punching, and melt-bonding. Linear elements having a continuous filament core with a protective sheath can be produced in strap, loop, and bar forms. Grids can be manufactured by cross laying and bonding together linear elements, or by drawing pre-punched sheets of polymer.

The manufacturing process and the form of the geotextile determine to a large extent the mechanical characteristics of the product and so performance data for generic polymers may be of little use to a designer. Performance tests must therefore be undertaken on each product, and standard tests methods must be applicable to the wide range of geotextile types. However the comparison of data from tests undertaken on different types of geotextile may not be straightforward.

3. Types of degradation

The polymer chains of the geotextile comprise a backbone of carbon atoms with occasional branches, side groups or additional atoms along the chain. Additives, to help facilitate manufacture and improve performance, and waste products, left over from the polymerization process, may also be present.

The chemical bonds can be broken or modified for example by oxidizing agents and ultra violet light, and additives may be leached from the bulk polymer: aggressive chemicals can be produced by micro-organisms. The rate of reaction is dependent upon the ambient temperature, the chemistry of the environment, and the presence of catalysts.

The installation process and the imposed in-service conditions can both result in the partial or total breakage of the structural elements, and distortion of the form of the geotextile. The combined effect of chemical attack and loading could accelerate certain types of degradation leading to premature failure of the plastics.

3.1. Chemical degradation

Oxidation. Oxidation is one of the principal forms of chemical degradation of polymers. All geotextiles contain anti-oxidants for stability in the short term, for the high temperature processing phase, and in the longer term, for lower temperature in-service conditions. The type and concentration of the anti-oxidants control the long-term thermal and oxidative performance, and so if they migrate by leaching with water or become saturated with oxygen then rapid degradation of the polymer may follow.

Hydrolysis. Hydrolysis is probably the principal form of chemical degradation of polyesters. The polymer chain is broken at the ester linkage by reaction with water, thereby reducing both the molecular weight and tensile strength of the chain. Considerable research effort has been directed at understanding the phenomenon. The rate of hydrolysis is dependent upon the molecular structure of the fibre and the ambient temperature. When the pH of the water is near neutral, the effects of temperature can be predicted well using the Arrhenius equation. But with highly alkaline environments, i.e. pH greater than 9, a different form of degradation occurs: with this form of degradation calcium ions seem to be more potent than sodium ions. High molecular weight polyesters should be selected to reduce the susceptibility to hydrolysis. Hydrolysis, using alkaline solutions, has been used to modify the surface of textile fibres to improve surface-related properties such as absorbency and wicking (or 'handle').

In-soil conditions. The pH of a soil is probably the most important factor governing the rate of chemical degradation of a buried geotextile. Surveys have shown that the natural soils in the U.K. have pH values ranging from about 4 to about 9. In acidic soils, such as peats, both humic and fulvic acids have been identified, whereas in alkaline soils calcium compounds are usually prevalent.

Generally, the polymers used for the manufacture of geotextiles have good resistance to acids and alkalis. It is only at the extreme range of pH's that the rate of oxidation and/or hydrolysis may compromise the integrity of the material. Such conditions may be present in waste dumps, in concrete, or in certain types of backfill such as minestone. There is also some evidence that the presence of polyvalent transition metal ions, such as iron, copper and manganese, can accelerate the rate of degradation of polyolefins.

The rate of chemical attack increases with increasing temperature. The mean temperature of the soils in the U.K. is around 10°C and this presents a relatively benign environment to buried geotextiles: higher temperatures are, however, encountered close to the ground surface and behind retaining walls. Temperatures can exceed 40°C in tropical climates, and even higher temperatures can be generated in waste tips.

3.2. Biological degradation

Evidence for biological degradation of geotextiles is sparse, but it is possible that organisms could consume additives, produce aggressive chemicals or even mutate to form polymer degrading variants. Geotextiles fibres may become clogged by organic growth or by the deposition of inorganic particles by micro-organisms.

3.3. Weathering

The degradation of plastics by the combined elements of outdoor weathering, i.e. light, air, moisture etc, is well documented.

In most applications, the geotextile will be buried in soil and will therefore only be exposed for a short time during construction works. There are a few applications where geotextiles are exposed, at least partially, during service: these include the use of meshes and grids for rock fall protection fences and for construction of gabion walls.

The resistance of geotextiles to natural weathering, in particular to sunlight, can be substantially improved by the addition of stabilizers.

3.4. Mechanical damage

Mechanical damage can be inflicted to a geotextile during its installation, particularly when the material is buried in soil, and by in-service loading. The type of damage will vary with the form of geotextile, and the effect on performance will depend upon the particular application, but it is clearly important to ensure that the function of the geotextile is not impaired by the installation process.

The degree of damage will be a function of the size, shape, and hardness of the soil particles and also the intensity of the compactive effort or, for vertical band drains, the forces applied during driving. Evaluation of the damage has been made by visual inspection, mechanical testing, hydraulic testing, and by thermal and chemical analysis to determine changes in the physical and chemical structure of the polymer.

4. Methods of assessment

A designer may only be concerned about the performance of a geotextile in a particular environment, however it would be difficult at the preliminary design stage to define, in quantitative terms, all the characteristics of an environment that could affect the performance of a geotextile. It would also usually be impractical at that stage to undertake a series of comparative tests on a range of geotextiles under conditions that simulated the in-service environment. These conditions would vary from site to site and some form of accelerated test would be required for long life applications.

4.1. Index tests

From necessity, index tests have been devised to determine the performance of a geotextile under 'artificial environments', i.e. ones that do not simulate typical in-service conditions. Most index tests for geotextiles are done in isolation and so the effects of the soil environment are removed. A series of tests is required to define the response of a geotextile to individual characteristics of a potential in-service environment, or simple combinations of these characteristics. Taken together these conditions may represent a rather severe environment, but this provides some degree of acceleration for long life applications and a margin of safety for short term uses.

An index test can be used to set a minimum performance standard and also to provide a means of ranking geotextiles in terms of their performance. Such tests may therefore be used to screen out those materials that may not give satisfactory service. The selection of the appropriate index tests depends upon the intended function of the geotextile, but it is essential that the designer is able to understand and define what constitutes a 'failure' for the proposed application.

4.2. Performance tests

The results of site trials can be used to calibrate or check the veracity of index tests, and so the procedure and acceptance limits for the index tests may be modified in the light of site experience. The potential range of soil environments is extremely wide and so the data from site trials are rather anecdotal. Moreover site trials are expensive and usually short term affairs. Nonetheless for some applications in-soil tests must be undertaken, preferably using the actual backfill from the site.

To bridge the gap between laboratory-based index tests and site practice, performance tests have been devised for some applications. The conditions imposed in these tests simulate, as close as possible, those occurring on site. Because the actual soil may not be known at the preliminary design stage, a range of standard soils, that cover those typically found in practice, may be used in these tests.

Unlike index tests the data from performance tests may be used directly, i.e. without modification, in design calculations.

Performance tests may be specified if the results of index tests indicate, for

Fig. 1. Micrograph of woven polyester after exposure to calcium oxide solution at pH 12.3 for 12 months at room temperature. While there was no significant reduction in strength or visible attack on the fibres, crystalline deposits occurred which could distort measurements of change of mass

example, that the performance of a geotextile may not be satisfactory in a highly alkaline soil.

4.3. Evaluation techniques

Observation. Specimens that have been exposed to weathering or recovered from test environments or sites can be examined to identify: changes in colour and texture; physical damage, such as the presence of holes and tears; chemical reactivity, such as crystal growth or dissolution; and the clogging of filter cloths. The nature of any change can be examined in detail using a scanning electron microscope. This is particularly effective in determining any surface attack on fibres, and the disruption of the structure of the fibres, but is not well suited for use in simple index testing. Fig. 1 shows a micrograph of a specimen of woven polyester taken after 12 months storage in an alkaline solution.

A damage factor can be determined by comparison of the areas that are 'undamaged' and 'damaged', for example the percentage area of holes or their size distribution. However, it is difficult to decide on what constitutes 'damaged' and 'undamaged' areas, and also what limiting ratio of these constitutes a 'failure' for a particular application. Thus observational techniques are normally used to provide qualitative rather than quantitative data, i.e. they show whether or not there has been any change. Such tests are by nature subjective and therefore open to operator interpretation.

Mass and dimensions. The mass and dimensions of a test specimen can be measured to check the stability of a material to a test or actual site environment. The tests are simple to perform and relatively small changes in mass or size can be determined. The test specimens should be conditioned in a controlled environment, both before exposure and following recovery from the test site, until they attain sensibly constant values: this may require alternate cycles of cleaning and drying. Unless the test regime models long-term usage, by for example using elevated temperatures, it is unlikely that any substantial changes in mass or size would be recorded for well established products. It may also be difficult to ascertain the importance of small changes in mass or size for many applications. For example small changes in mass or size due to the absorption of water may have little or no effect on the ability of a geotextile to function as a separator or reinforcement. Thus the diagnostic value of these simple tests may be limited.

Tensile tests. Almost all forms of degradation will affect the mechanical properties of a geotextile, and in most applications this will have some bearing on the ability of the material to function properly.

The degree of degradation could be characterized by the ratio of the rupture strengths of as-supplied and exposed tests specimens. However a designer may be more concerned about changes in stiffness than strength. Changes in the elongation at rupture may also be a more sensitive measure of degradation than rupture strength, particularly when the material shows a yield point or is flawed. It should be appreciated that tensile tests are index tests and careful interpretation of the data is required.

Statistical checks are used to assess whether or not there has been any significant degradation, and so at least five repeat tests are required to determine the mean and scatter of the data. More than five tests may be required with certain types of material, such as heavy-duty non-woven geotextiles, to adequately define the statistical values particularly where the change in the measured parameter, such as rupture strength, is less than 10 per cent.

5. Codes of practice, standards and methods of test

The British Standards Institution (BSI) commissioned the production of a Code of Practice (BSI,1992) to provide guidance to users of geotextiles on the requirements and methods of testing for durability. This document was prepared, under the auspices of Technical Committee M35, by D. I. Cook, of the British Textile Technology Group, with contributions from J. H. Green-

wood, of ERA Technology Ltd, and J. E. Uprichard, then of the Lambeg Industrial Research Organisation. A revised draft that included contributions from the European Technical Standards Committee (CEN.TC189) is currently (1992) being prepared for publication: sections of this paper were drawn from the text of that draft.

Development of standard tests for geotextiles is continuing, with most of the activity taking place in Europe and America. The BSI, through TCM-35, have prepared eight parts of BS 6906 (1987 to 1991), but are now restricted from publishing further test methods under the 'standstill' procedure imposed by CEN. The CEN committee (CEN.TC189) has representatives of most European Countries, and is currently assessing standard test procedures for geotextiles. There is considerable pressure to produce the necessary documentation so that members of the European Community can conform to the Construction Products Directive on the introduction of the Single Market. The American Society for Testing Materials (ASTM) has issued a number of standards on geotextiles and geomembranes, and these have now been produced together under one cover (ASTM,1991). Other standards have also been issued by the Geosynthetic Research Institute of Drexel University, and also by the federal agencies.

The BSI holds the secretariat of the International Standards Organisation (ISO) technical committee 38/SC21; this has produced a few international standards and has also sent observers to the CEN meetings. There should, through the activities of the ISO, be a general harmonisation of European and American standards along with those from other countries that have a relatively high use of geotextiles, such as Japan.

5.1. Chemical degradation

Oxidation. The testing standards committees is evaluating a test method based on initial work completed in the Netherlands and reported by Wisse *et al.* (1990). The test was originally devised to determine the long term stability of geotextiles either immersed in water or exposed to natural weathering. Test data show that the oxidation of polypropylene geotextiles proceeds more rapidly in air than in water. The method is based on the determination of the time to embrittlement of test specimens exposed to high temperatures, somewhere between 120 and 140°C. The test specimens can be immersed in hot water prior to exposure to high temperatures so that the tendency for leaching out of additives, such as anti-oxidants, can be assessed.

Hydrolysis. Testing standard committees are evaluating index tests for screening-out geotextiles that are sensitive to hydrolysis. Such tests may be invoked when for example polyester geotextiles are being considered for use in alkaline environments. The tests will probably include the exposure of geotextile specimens to (i) a high temperature environment to which steam, under pressure, is introduced and (ii) an alkaline solution.

In-soil conditions. The pH of natural soils in the U.K. normally falls within the range of 4 to 9, and the mean ambient in-service temperature for most applications is about 10°C with the maximum temperature rarely exceeding 20°C. Most soilspresent rather a benign environment to geotextiles, and so even in the long term most geotextiles will not suffer any substantial degradation due to burial. However geotextiles may be exposed to more testing environments when used in contaminated land sites, such as waste dumps.

It has been proposed to devise index tests to identify those geotextiles that are sensitive to particularly acidic and/or alkaline environments. These materials would not necessarily be rejected for use in such environments, but would be subjected to more specific tests to evaluate their likely performance for the particular application. The index test would involve the immersion of the test specimens in various fluids at a temperature of 50°C for a period of 28 days. These fluids would test the stability of the geotextile to acid and alkaline environments. The acids under consideration include acetic acid, an easily obtainable organic acid, and parahydroxybenzoic acid, a principal component of peaty soils, further details have been given by Billing *et al.* (1990). The presence of multi-valent transition metals is known to accelerate the oxidation of some polymers, and so the use of an iron salt solution is also being considered.

Performance tests could be undertaken if the geotextile showed appreciable deterioration in the index tests or was likely to be exposed to exotic chemical compounds; these include high concentrations of particular transition or heavy-metals; weedkillers or fertilisers, particularly ammonium compounds; organic solvents; anti-freeze compounds; and gaseous pollutants, such as the oxides of sulphur and nitrogen. The procedure and apparatus for the performance tests would be similar to the index tests. Initial tests may also be undertaken using a storage temperature of 50°C, but if degradation is observed then further tests may be undertaken using different storage temperatures to assess the likely degradation at the in-service temperature. This approach is similar to that advised by the U.S. Environmental Protection Agency for evaluating the stability of membranes, USEPA (1986).

5.2. Biological degradation

There are well-established procedures for exposing materials to specific strains of bacteria and fungi grown in the laboratory. However it is more useful to expose the geotextile specimen to a biologically-active environment that better simulates in-soil conditions, and an in-soil burial test, such as described in section 2 of BS 6085 (1992), is simple to set up. In such tests, the specimens are buried in a soil that supports a large and varied population of micro-organisms. It is not necessary to specify a standard soil but the key criterion must be met that a control sample of cotton cloth is degraded within seven days. To achieve this normally requires the use of topsoils and composts, as recommended in BS 6085, and coarse-grained soils are generally not suitable. The moisture content of the soil must be controlled: BS 6085 suggests a moisture content of between 20 and 30 per cent, but the test method given in DIN 53739 (1984) specifies a moisture content equal to 60 per cent of the maximum holding capacity. The soil environment is maintained at a temperature of 28±1°C to ensure rapid propagation of the micro-organisms. The soil from a particular site could be used, but it may be necessary to remove all the larger particles to ensure that the key criterion is met. A test method based on this procedure is to be proposed to the CEN standards committee TC-189.

A further test is required to assess the blocking or clogging of filter fabrics by biological activity. Tests have been undertaken to evaluate the change in the rate of flow of biologically active fluids and leachates through geotextiles. Such tests have normally been run for a period of 12 months. Accelerated chemical degradation may be associated with the clogging of filters by iron compounds deposited by micro-organisms.

5.3. Weathering

Tests are required to assess (i) the performance of those geotextiles that are only exposed to natural weathering at the construction stage, and (ii) the performance of the geotextiles that are at least partially exposed throughout their intended service life. Whereas a simple index test should cater for the first category, it will be necessary to undertake accelerated ageing tests and confirmatory natural weathering trials for the second.

Accelerated tests are undertaken using high intensity light sources, such as the Xenon arc and fluorescent lamps, whose spectra, covering the ultraviolet and visible range, are similar to those of natural daylight. Procedures for performing accelerated laboratory tests were issued by ISO in 1981; these procedures are currently being reviewed. A number of comparative studies of the results of accelerated and natural weathering tests have been completed. The rates have been found to vary widely from one material to another so that no general rules have been established to enable the rates to be estimated for new materials. Thus extensive comparative tests are required for all materials that are exposed to sunlight during their in-service life.

5.4. Mechanical damage

Samples of geotextiles have been extracted, some time after their installation, from sites and examined to determine the forms and degree of degradation. The results of studies completed in France and the U.S.A. have been reported

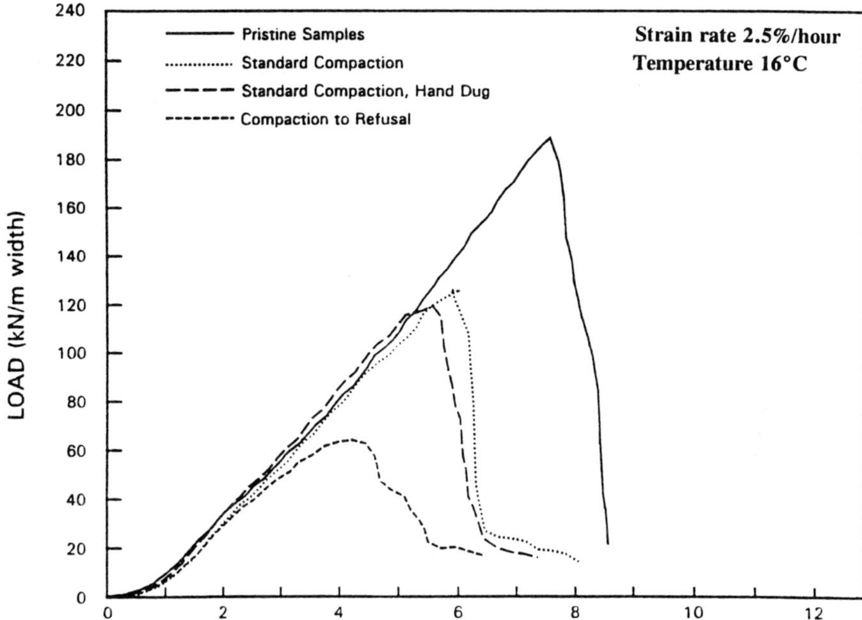

Fig. 2. Relation between load and strain for Polyester P3

by Leclercq *et al.* (1990), and Koerner and Koerner (1990) respectively. The general conclusion of this work, and of similar studies completed in the U.K., was that for most applications the installation process was likely to be the most important source of damage to geotextiles.

The site-specific studies did not distinguish between damage due to the installation process and due to in-service conditions: moreover, because the type of backfill and the level of compactive effort varied, comparison of the data from different sites was problematic. It also proved time-consuming and difficult to extract the samples from sites without inflicting further damage. For these reasons, work in the U.K. concentrated on establishing a test procedure that simulated the installation process on typical construction sites.

The procedure was developed by TRL and ERA as a part of a three year multi-client study into the durability of geotextiles. Details of the test procedure and preliminary results have been provided by Watts and Brady (1990). Some of the data are reproduced in Fig. 2 and Table 1. Further details of the tests and the results of the validation trials were provided by Greenwood and Brady (1991). The main conclusions of the work were that

(a) the extent of the damage was dependent upon the coarseness of the backfill and the intensity of the compactive effort

(b) the damage was characterized by a reduction in rupture strength but the stiffness of the geotextiles was largely unaffected
(c) qualitative measures of the level of damage, i.e. the number of holes, tears in the materials, were too dependent upon operator interpretation to be of use for a standard method.

Further work is required to more closely define the test method, and to determine the manner in which tensile tests should be undertaken on severely damaged material. It is likely that the test procedure will be proposed as a standard means of defining installation damage.

6. Conclusions

The evidence from the tests show that most geotextiles are extremely durable. Nonetheless, questions of durability are frequently raised by specifiers of geotextiles, and engineers require guidance on both the testing and the interpretation of the test data: this is particularly true for long-term applications in the more demanding environments.

Table 1. Results from tensile test on 3 woven geotextiles and one geogrid, to assess the reduction in strength due to installation damage

FABRIC TYPE	TENSILE STRENGTH kN/m	STRAIN AT MAX LOAD %	COMMENTS
P1	190	12	Mean of 5 tests on pristine specimens
	177	11	Standard Compaction
	183	12	Standard Compaction, Hand Dug
	122	8	Compacted to Refusal
P2	46	11	Mean of 5 tests on pristine specimens
	44	9	Standard Compaction
	39	10	Standard Compaction, Hand Dug
	21	6	Compacted to Refusal
P3	188	8	Mean of 5 tests on pristine specimens
	126	6	Standard Compaction
	119	6	Standard Compaction, Hand Dug
	65	5	Compacted to Refusal
GRID	54	12	Mean of 4 tests on pristine specimens
	50	10	Standard Compaction
	50	10	Standard Compaction, Hand Dug
	47	9	Compacted to Refusal

All tests performed on 1m wide specimens at a rate of 2.5% strain per hour. Temperature 16°C.

Brief descriptions of the various agents of degradation and the methods used to characterize the degree of degradation are provided in this paper. It also provides details of the activities of various committees charged with defining test methods for geotextiles. Although there already exists a broad catalogue of test methods, further work is required to define and validate new tests to encompass the full range of applications. Laboratory-based tests for durability have not yet been widely used, and any proposed new test method requires an extensive validation exercise. The data from accelerated weathering tests must be compared to in-service conditions to define the level of acceleration.

As in other fields of engineering, the confidence placed on the durability of geotextiles will develop gradually as the results of long-term tests become available, and as the mechanisms of degradation are better understood.

7. Acknowledgements

The work described in this paper forms part of a multi-client funded research programme co-ordinated by ERA Technology. The work was part funded by DTI and the Department of Transport. The paper is published by permission of Mr J. E.Hall and Dr A. Gallagher (Customers of TRL at the Department of Transport), the Chief Executive of TRL, and the Directors of ERA Technology.

8. References

AMERICAN SOCIETY FOR TESTING AND MATERIALS (1991). *ASTM Standards on Geosynthetics*, Second Edition. Publication Code Number 0343509138.

BILLING J.W., GREENWOOD J.H., SMALL G.D. (1990). Chemical and mechanical durability of geotextiles. *Fourth Int. Conf. on Geotextiles, Geomembranes and Related Products.* The Hague. Balkema. Rotterdam. Vol 2, pp 621-626.

BRITISH STANDARDS INSTITUTION (1992). BS6085. *Methods for determination of the resistance of geotextiles to microbiological deterioration*. BSI.

BRITISH STANDARDS INSTITUTION (1992). PD6533 *Methods for assessing the durability of geotextiles - an interim code of practice*. BSI.

BRITISH STANDARDS INSTITUTION (1987-1991). BS6906. *Methods of test for geotextiles.* BSI.

DEUTCHES INSTITUTE FUR NORMUNG EV (1984). DIN 53739. *Testing of plastics, influence of fungi and bacteria, visual evaluation, changes in mass of physical properties.* DIN.

GREENWOOD J.H. and BRADY K.C. (1991). Geotextiles in aggressive soils. *International Conference : Polymers in extreme environments.* University of Nottingham. The Plastics and Rubber Institute. Paper 10.

INTERNATIONAL STANDARDIZATION ORGANIZATION (1981). ISO 4892. *Plastics - methods of exposure to laboratory light sources.*

KOERNER G.R. and KOERNER R.M. (1990). The installation of survivability of geotextiles and geogrids. *Fourth Int. Conf. on Geotextiles, Geomembranes and Related Products.* The Hague. Balkema. Rotterdam. Vol 2, pp 597-602.

LECLERCQ B., SCHAEFFNER M., DELMAS P., BLIVET J.C. and MATICHARD Y. (1990). Durability of geotextiles : Pragmatic approach used in France. *Fourth Int. Conf. on Geotextiles, Geomembranes and Related Products.* The Hague. Balkema. Rotterdam. Vol 2, pp 679-684.

UNITED STATES ENVIRONMENTAL PROTECTION AGENCY (1986). Method 9090. Compatibility test for waste and membrane liners. In *test methods for evaluating solid waste.* Vol 1A: Laboratory Manual, Physical/Chemical Methods, 3rd Edition., SW-846, USEPA, Washington DC, USA.

WATTS G.R.A. and BRADY K.C. (1990). Site damage trials on geotextiles. *Fourth Int. Conf. on Geotextiles, Geomembranes and Related Products.* The Hague. Balkema. Rotterdam. Vol 2, pp 603-607.

WISSE J.D.W., BROOS C.J.M. and BOES W.H. (1990). Evaluation of the life expectancy of polypropylene geotextiles used in bottom protection structures around the Oster Schelde storm surge barrier, a case study. *Fourth Int. Conf. on Geotextiles, Geomembranes and Related Products.* The Hague. Balkema. Rotterdam. Vol 2, pp 697-702.

Discussion

Reporter : R. Woods, Surrey University, UK

C. LAWSON, Exxon Chemical Geopolymers
Dr Cazzuffi says that thermally bonded geotextiles are different to needle punch - is this because of their different structure or their different initial hydraulic properties?

D. CAZZUFFI, ENEL
I am not sure. Leachate is not the same as water. In the laboratory we were only looking at the effects of leachate. In practice it is necessary to also consider the characteristics of soils and leachate and also at leachate-geotextile-soil interaction.

CHAIRMAN
Did Dr Cazzuffi look at what was growing in the geotextiles? Did he analyse the clogging elements?

D. CAZZUFFI, ENEL
Yes. Some chemical/biological examinations were carried out by specialists: the tests we carried out were not specifically to look at growth but purely to look at the effects of clogging. Koerner at Drexel has done work in this area.

K. ROWE, University of Western Ontario
How did you maintain anaerobic conditions in the test? What changes in characteristics did you see in the leachate during the test and how often was the leachate changed?

D. CAZZUFFI, ENEL
The specimens were in sealed controlled conditions to keep the leachate anaerobic. Results from the tests after 3 and 5 months: we used another

leachate sample. Details are in the paper and the differences are recorded.

CHAIRMAN
For those of us who have not had the strength of character to carry out research with leachate from landfills, could the growth and clogging be inhibited by surface-dressing the geotextile or by adding chemicals to the polymers?

C. LAWSON EXXON, Chemical Geopolymers
Some work at Drexel some years ago showed that the inclusion of biocides in geotextiles was not successful.

CHAIRMAN
Are there any comments on the 'T and O' findings that some loss of fines is acceptable, as this is a different approach?

C. LAWSON, Exxon Chemical Geopolymers
We have designed filters for projects where fines were intended to pass through geotextile. However, one must be careful, as the fines will tend to clog the drainage layer. The drainage layer must therefore be designed to accommodate these fines or you need a procedure for back-flushing.

CHAIRMAN
Yes, that is okay if you have a drainage layer. River applications appear to be more problematic due to the possible two-way flow situation.

B. MYLES, Consultant
The type of design being considered is frowned upon in the USA. With uniformly graded soils in coastal protection situations and trafficked waterways, soil loss can be large even in relatively still conditions. We need to be very cautious.

CHAIRMAN
Designers may not have a lot of choice on a very large project; the ground investigation may not reveal all the characteristics of the protected soils.

S. CORBET, G. Maunsell & Partners
Revetments are of particular concern in a situation of migrating fines. Instability may occur with the revetment eventually collapsing; therefore to allow migration is not wise.

CHAIRMAN
Yes we do have a situation for concern; once the soil fines are moving under

the hydraulic gradients in the scheme, where will they go? If they clog the filter the whole protection system could lift off or if they pass through we will have subsidence.

H. WICHERN, T and O
In Holland we were not designing it, more accepting the facts - based on Dutch experience 10-16% fines can be lost from the thin layer (5 cm) beneath the filter but we still have stable sub-strats.

D. AYRES, Consultant
Considering railway experience we have seen clay size particles pumping up through filters to a height of up to 50 mm or more above the geotextile but not much higher. The geotextile seems to slow down the penetration of fines into the track ballast but does not stop it completely. Drying and wetting may alter things over a period to time.

CHAIRMAN
It would be intriguing to develop an accelerated test mechanism but our present understanding of the way soil fines move is too imperfect to work out what factors we need to increase - hydraulic gradient or the frequency of cyclic load application. We seem to need some more fundamental work on the problems.

H. WICHERN, T and O
One of the findings in the Dutch tests was the effect of iron oxides on polypropylene. Could Dr Greenwood comment?

J. GREENWOOD, ERA
A separate test is needed to assess the catalytic effect of iron in the oxidation of polymers, in particular of polypropylene. The test will be carried out in an accelerated environment.

K. ROWE, University of Western Ontario
Dr Greenwood is going to carry out index tests using substances with pH values less than 3.4 and greater than 12.5. However, as the typical range of pH is 4 to 9, why use such an extreme range?

J. GREENWOOD, ERA
The problem is one of acceleration. We are trying to predict behaviour after 20 to 120 years. Although we can accelerate degredation effects by raising the temperatures, in some circumstances it will also be necessary to increase the concentration. When we have more practical experience we may be able to choose more appropriate parameters.

The way forward: the consulting engineer's view

T. S. INGOLD, Consulting Engineer, St Albans, UK

1. Introduction

Compared with graded aggregate filters and drains those incorporating geotextiles claim to offer advantages both in terms of economy and performance. Over the past two decades several millions of square metres of geotextiles have been installed as filters and yet analysis of published design rules show some glaring inconsistencies. Since the rate of reported failures of geotextile filters is low, a reasonable conclusion might be either that the design rules are conservative or, taking a jaundiced view, that geotextiles always work when they are not needed. With the exception of the small minority of research based on field monitoring of performance the majority of design rules tend to emanate from analytical studies or laboratory based testing. Even at the philosophical level this leads to inconsistencies with one group of researchers maintaining that the geotextile filter is the controlling factor, Corbett and Farrar (1990), whilst others argue that compared to the geotextile the soil itself has at least the same, if not a greater, influence on performance, Dierickx and van der Sluys (1990).

2. Inconsistencies in design methods

The requirements for a geotextile filter are essentially the same as those for a graded granular filter in so far as the pore sizes of the filter should be small enough to prevent an unacceptable migration of base soil particles yet not so small as to impeded the flow of liquid from the base soil. In designing graded granular filters the retention and permeability requirements are most commonly met by ensuring that there is a certain required degree of equality between the particle size distribution curves of the filter medium and the base soil.

Design methods for geotextile filters fall into one or the other of two broad categories. The first of these involves the use of a set of geometric design rules based on considerations of the pore size, or sizes, of the geotextile and the particle size distribution of the base soil. Some of the early geotextile design rules reflected those used for graded granular filters and involved obtaining the requisite equality between the pore size distribution of the geotextile and the particle size distribution of the base soil, Rankilor (1981). It is interesting to note that some researchers, Fischer *et al.* (1990), Holtz (1990), are now seeking to reintroduce this approach since, they argue, it simultaneously addresses the issues of retention and permeability whilst not differentiating between woven and nonwoven geotextiles.

Current design rules are reviewed in detail elsewhere, Giroud (1988), Fischer *et al.* (loc cit), Heerten (1986). To meet the retention criterion, or criteria, most require a certain minimum ratio between a selected geotextile pore size and a selected base soil particle size. Often this ratio will be varied according to the structure of the geotextile filter and quite commonly there is differentiation between woven and nonwoven geotextiles. One of the best known examples is that due to Ogink (1975).

$O_{90}/d_{90} \leq 1.0$ for wovens
$O_{90}/d_{90} \leq 1.8$ for nonwovens

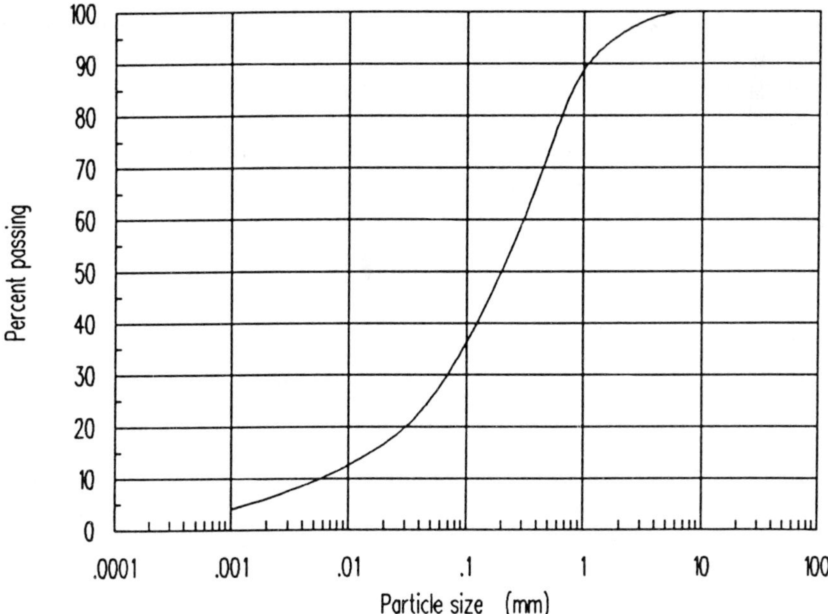

Fig. 1. Base soil particle size distribution

To meet the permeability criterion, or criteria, there is generally a requirement for the permeability of the geotextile, k_g, to bear a certain ratio to the permeability of the soil, k_s. Depending upon the particular set of design rules being used this ratio varies between 0.1, Giroud (loc cit) and something in excess of 100, Comité Français (1986).

The other category of design methods encompasses laboratory testing of soil-geotextile systems. Perhaps the most well known of these are the GR (gradient ratio) test developed by the US Army Corps of Engineers, Haliburton and Wood (1982), the LTC (long term column) test developed at the Geosynthetics Research Institute, Koerner and Ko (1982), and the HCR (hydraulic conductivity ratio) test, Williams and Abouzakhm (1988). Essentially all these tests seek to monitor how the system flow rate, or hydraulic gradient, changes with time. In turn this allows an assessment of the geotextile from the point of view of retention and permeability.

It is beyond the scope of this paper to enter into a detailed consideration of the pros and cons of various filter design rules or design methods based on systems testing, however, it is enlightening to consider the following problem. Fig. 1 shows the particle size distribution of a base soil to be filtered by a geotextile. The soil is a silty sand with d_{85}=0.8 mm with a coefficient of uniformity of 67. The geotextile filter used was a 204 g/m^2 woven polypropylene fabric with O_{95}=210 μm. The suitability of the geotextile was assessed using various geometric filter design rules and assessments of the results from the GR, LTC and HCR systems tests. For the geometric design rules consideration is limited here to retention, however, further details are given by Williams and Luettich (1990).

The geometric design rules considered are those by Giroud (1982), the Federal Highways Adminstration (FHWA), Christopher and Holtz (1985), the Comité Français des Géotextiles et Géomembranes (loc cit) and Working Group 14 of the German Society for Soil Mechanics and Foundation Engin-

Table 1. Maximum permitted O_{95} from various design rules

Design Rule	Maximum O_{95} (μm)
Actual	210
Giroud	110
FHWA	800
CFGG	1000
GSSMFE	204

Table 2. Failure predicted by various design methods

Design Method	Failure Predicted
Giroud	Yes
FHWA	No
CFGG	No
GSSMFE	Yes
GR	No
LTC	No
HCR	Yes

eering (GSSMFE), Heerten (loc cit). The O_{95} of the geotextile actually used was 210 μm and Table 1 shows the maximum value of O_{95} permitted by the various design rules. As a corollary to this, Table 2 shows whether or not these design rules, and the assessments made from various systems tests, predicted failure.

The geotextile actually used is reported as having failed in the field due to clogging leading to inadequate flow capacity, Williams and Luettich (loc cit). Of the seven design methods cited above only 43% predicted the failure; the prediction by the GSSMFE design rules was marginal. Although the predictions from the design rules by Giroud and the GSSMFE were correct in as much as they indicated failure, the predicted failure mechanism was by piping whereas the reported failure mechanism involved clogging. In this instance interpretation of the HCR test results accurately predicted failure by clogging leading to inadequate system permeability.

The above comparisons are not intended to endorse any of the design methods cited. It might well be that a completely different ranking of accuracy of prediction would be achieved using a different soil and a different geotextile. However, the above inconsistencies might lead to problems for the designer arguing against design liability in a court of law.

3. Inconsistencies in pore size determinations

Reference to the proceedings of the fourth international conference held in the Hague reveals a range of geotextile pore sizes including the apparent opening size (AOS), the filtration opening size (FOS), D_W, O_{95}, O_{95r}, D_{95}, O_{90}, O_{50}, O_{15}, and O_f. A cull of the literature appears to reveal that the AOS, O_{95}

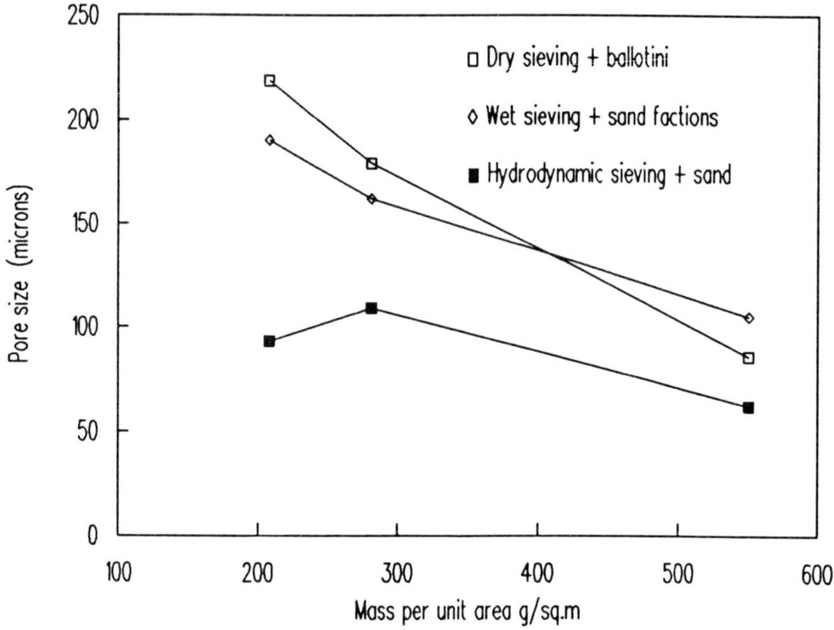

Fig. 2. Variation of nonwoven pore size with test method

and O_{95r} are one and the same thing derived from dry sieving. Likewise FOS and D_{95} appear to be one and the same thing derived from hydrodynamic sieving. D_W stands alone and is derived from wet sieving, whilst O_{90} is derived from dry sieving and is associated with the d_{90} particle size in most European filtration design rules, and is not to be confused with O_{95} which is associated with d_{85} in most US design rules. The pores sizes O_{50}, and O_{15} are used in a minority of cases and are therefore considered no further. The definition of O_f appears to vary from one author to another. The test methods cited include wet sieving, dry sieving, hydrodynamic sieving, mineral aggregate or glass beads, and these are only the index tests.

Confused? Some clarification is given in the proceedings of the third international conference held in Vienna. 'In the recommendations of WG14 GSSMFE the filter rules refer to the effective opening size D_W. D_W is determined by the wet sieving analysis of the Federal Institute for Hydraulic Research and Coastal Engineering. This testing method is also laid down in a Swiss standard (SN640550), however, the effective opening size designated is O_W. The effective opening size D_W corresponds to undersize amounting to 10% of test soil and 90% oversize. Therefore, according to the declarations for symbols and dimensions of this conference, D_W=O_W=O_{90}. But it must be strictly noted that the defined O_{90} resulting from the wet sieving test of the

Franzius Institute for Hydraulic Research and Coastal Engineering cannot be treated as equivalent to O_{90} values resulting from other test methods. Each O_{90} value has its specific test definition. The following filter rules can therefore only be applied with O_{90} values resulting from the mentioned wet sieving method or otherwise a O_{90}/d_{90} relationship for different test methods has to be considered.' (Heerten (loc cit)). This statement essentially makes the very important point that if a particular set of geometric design rules are to be used then they must be used with geotextile pore sizes determined by the test method associated with the design method.

That great care should be taken in relating filtration rules to appropriately measured pore sizes is reflected in Fig. 2, based on Gourc and Faure (1990), which shows how measured pore size varies with mass per unit area for a nonwoven needlepunched geotextile and three different test methods. Although there are slight variations, the magnitudes of the pore sizes measured by hydrodynamic sieving are approximately half those measured using dry sieving with ballotini or wet sieving with sand factions. Even when only one test method is used the results obtained should be treated with caution. This is exemplified by Fig. 3, based on Rollin *et al.* (1990), which shows how pore size, measured by hydrodynamic sieving, varies with mass per unit area for nonwoven needlepunched geotextiles produced by different methods.

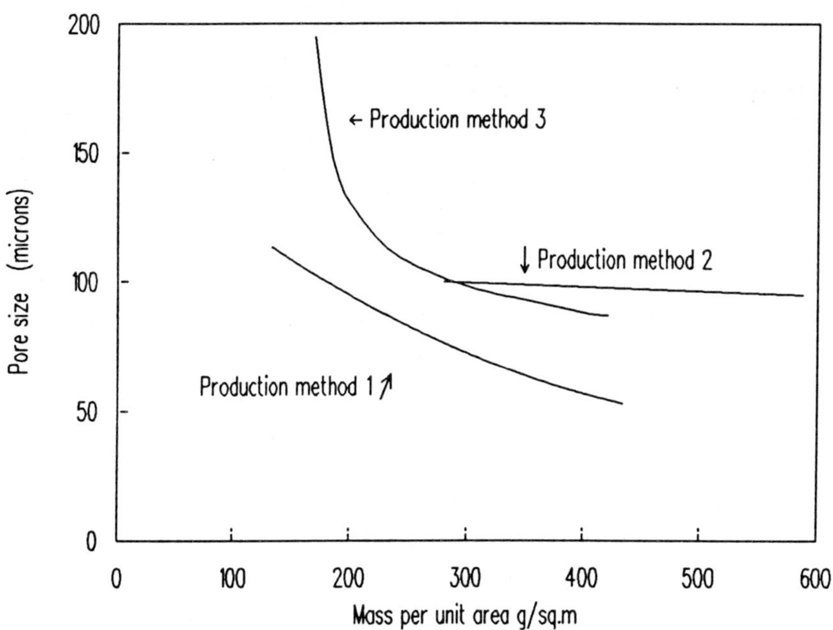

Fig. 3. Variation of nonwoven pore size with production method

The variations in measured pore size described above are not peculiar to nonwoven geotextiles. Table 3 gives results derived from inter-laboratory trials on a 130 g/m^2 woven polypropylene tape fabric. As can be seen these results reflect not only a variation in measured pore size due to test method but also, for a given test method, they show the influence of operator variable. For a given test method Table 3 indicates a factor of almost 2 between the maximum and minimum values. Comparing the average pore size values from dry sieving with those from hydrodynamic sieving shows the average hydrodynamic value to about 1.5 times larger than the average dry sieving value.

4. The need for rationalisation

That there as are as many different answers to a filtration problem as there are design methods leads to the conclusion that there is a need for rationalisation. This requires normalisation of both test methods and design methods. To a certain extent the normalisation of test methods is a comparatively simple process which revolves around the willingness of all parties to agree upon what the standard test methods are to be. Since geotextiles, by their nature, are variable, and their perceived properties are a function of the test method used to measure them it is a fools errand to chase a definitive test method which a gives a definitive measure of the true pore size. The truth of this is revealed by a study of the history of the measurement of soil shear strength which occupied academics for over three decades.

Within western Europe the normalisation of test methods for the determination of pore size will soon be completed by CEN Technical Committee 189. The likely outcome is the use of a range of test methods, such as dry sieving,

Table 3. Variations in measured woven pore size

Laboratory	**Test method and pore size (μm)**	
Name	**Dry Sieving**	**Hydrodynamic Sieving**
A	-	216
B	-	392
C	91	-
D	151	-
E	172	-

wet sieving and hydrodynamic sieving, but with the use of each test method confined to a certain predetermined range of pore size. Although the measured pore size may not be definitive at least it should be consistent. One step which would aid unification on a worldwide basis is a decision as to what is the representative pore size. At present the Europeans tend to use O_{90} whereas the North Americans seem to prefer O_{95}.

Similar observations apply to the measurement of geotextile permeability. Is this to be couched in terms of permittivity, water permeability measured at constant head, permeability measured at a standard hydraulic gradient or permeability measured at a constant flow velocity.

In determining standard methods of test a very clear demarcation needs to be made between test methods intended to provide index values of properties and tests methods to provide hydraulic properties to be used in design. For straightforward design problems, where the ramifications of failure are small, it should be possible to devise simple and safe design methods based upon index values. If there is to be a worldwide unification of such simple test methods then there is a need to agree on the representative particle size of the base soil as either d_{85} or d_{90}.

For more onerous design problems each solution is likely to be unique, however, the solution needs to be based upon the use of standard test methods for determination of design values of hydraulic properties and/or the use of standard systems tests. The logic here is that if all parties work to standard test methods then the pool of practical knowledge amassed from monitoring failures and successes can be compared to the same base. A classic parallel in soil mechanics is the use of the standard penetration test (SPT). Whilst the test itself is crude it has the advantage of being cheap, universally practised, and is backed by an extensive pool of experience which forms the basis of a wide range of practical design methods.

Although the development of design rules has exercised the minds of academics and practitioners over the past twenty years there are still those who believe that definitive analytical methods can be developed. For those of this bent it would do well to reflect on the words of Terzaghi and Peck which, although quoted close on fifty years ago, ring as true today as they did then: 'Soil mechanics originated several decades ago under the pressure of necessity. As the practical problems involving soils broadened in scope, the inadequacy of the scientific tools available for coping with them became increasingly apparent. Efforts to remedy the situation started almost simultaneously in the United States and in Europe, and within a short period they produced an impressive array of useful information. The initial successes in this field of applied science were so encouraging that a new branch of structural analysis appeared to be in the making. As a consequence, the extent and profundity of theoretical investigations increased rapidly, and experimental methods were developed to a high degree of refinement. Without the results of these painstaking investigations a rational approach to the problems of earthwork engineering could not have been attempted. Unfortunately, the research activities in soil mechanics had one undesirable psychological effect. They diverted the attention of many investigators and teachers from the manifold limitations imposed by nature on the applications of mathematics to problems in earthwork engineering.'

5. The way forward

Most of the foundations for practical methods of design and testing have been laid. However, if these are to be developed for use by the geotechnical engineering community at large then there is a need to classify the drainage and filtration problems to be tackled to ensure that the method of design addresses all the controlling factors.

An obvious initial classification might be with regard to the magnitude of the works and the ramifications of failure. Clearly the consequences of failure of a 50 m run of geotextile wrapped drain are very different from those for chimney drain in an earth dam or a prefabricated wall drain behind a major bridge abutment. If this is accepted then the margins of safety need to reflect this differential.

As a design aid the problem must be classified according to hydraulic gradient. Clearly there is a need to differentiate between steady state and alternating flow but additionally there is a need to differentiate between the inclination of the drain, be it vertical, horizontal or inclined, and the inclination of the hydraulic gradient. This comment applies to analytical design methods and particularly soil-geotextile systems testing.

Simulation or systems tests offer huge potential if correctly exploited. Sight can be lost of the effect of hydraulic gradient on the prevailing effective stress regime which in turn affects how a particular geotextile is perceived to perform. For example in a saturated soil with a static ground water table coincident with ground level the hydraulic gradient is zero and the vertical effective stress at a depth z is $z(\gamma-\gamma_w)$. If there is subsequently vertically downwards infiltration under gravitational flow, the hydraulic gradient i assumes a value of unity and the vertical effective stress at a depth z becomes $z(\gamma-\gamma_w+i\gamma_w)\approx z\gamma$. The vertical effective stress, and therefore stability, is increased. On the other hand if there is upwards vertical flow under a hydraulic gradient of unity the vertical effective stress at a depth z becomes $z(\gamma-\gamma_w-i\gamma_w)\approx 0$, in which case the vertical effective stress, and therefore stability, decrease. Although the use of high unidirectional hydraulic gradients in systems or simulation tests may reduce test duration times it is preferable that they be applied in a direction and magnitude which more accurately simulates realistic effective stress regimes.

Other major factors which should be classified are the base soil and time.

Several aspects of the base soil require classification, including its particle size distribution, effective cohesion and chemistry. The latter is important from several points of view including chemical attack of the geotextile, geotextile clogging caused by the precipitation of chemicals dissolved in the groundwater and geotextile clogging caused by micro-organisms.

The classification of design life is important since where long service lives are required due attention must be paid to durability of the material of the geotextile as well the factors which could cause long term clogging. At the other end of the timescale, and for all applications irrespective of design life, the issue of survivability must be addressed. Although the research related to separators, geotextiles exhumed by GRI shortly after installation showed that damage caused during installation resulted in up to 120 holes per square metre with a corresponding loss in tear strength of 88%, Koerner and Koerner (1990). Consequently having carefully designed for a pore size of 250 μm, or whatever, it is important to ensure that the geotextile has adequate mechanical properties to prevent this pore size being ripped to something many times larger during installation.

6. Conclusions

For geotextiles to become fully accepted as a civil engineering construction material on par with steel or concrete it is necessary to marshall test and design methods into appropriate standards or codes of practice. This is being done successfully in other fields of application, such as soil reinforcement. Recommendations for drainage and filtration currently available in Europe are, unfortunately, often contradictory.

Standards for index test methods have been established in the USA and in the near future European test methods will be defined through CEN. It is possible that CEN will also introduce classification systems for geotextiles including geotextiles used as filters and drains in low risk applications. As a part of any such classification system it will be necessary to define a representative pore size and it will be interesting to see whether this will be O_{90}, in line with the majority of existing European design rules, or O_{95} in line with established US test methods and design rules. Since existing European design rules will need to be revised to take account of the likely reallocation of pore size test methods it would seem sensible to make the additional effort to settle on O_{95} as the representative pore size.

For more demanding filtration and drainage applications analytical designs based on index test properties may not be adequate. By definition index testing does not account for site-specific parameters and consequently in more critical applications the designer must turn to other types of testing. In ascending order of relevance these may be classified as laboratory performance tests, systems tests, large scale model tests and finally prototype or full scale field tests. Whatever classification is finally used it is important that there be uniformity of purpose particularly with respect to modelling of hydraulic gradients and effective stress regimes.

In general manufacturers do not guarantee their products or the performance of their products in a given application. In applications which unfortunately fail this raises the issue of design liability. As has been demonstrated in this paper the present plethora of design methods and test methods leads to an array of often conflicting answers to the same problem. This would be likely to give rise to a very complicated situation where design liability has to be settled in a court of law.

7. References

CFGG (1986) *Recommendations pour l'emploi des géotextiles dans les systèmes de drainage et de filtration*. Comité Français des Géotexties et Géomembranes., Paris, 23pp.

CHRISTOPHER B.R. and HOLTZ R.D. (1985) *Geotextile engineering manual.* Federal Highways Administration, Report No. FHWA-TS-86/203, 1044p.

CORBETT S.P. and FARRAR D.M. (1990) Comparative field trials of fin drains. *Proc IV Int. Conf. on Geotextiles, Geomembranes & Related Products*, The Hague, Vol. 1, 367.

DIERICKX W. and VAN DER SLUYS L. (1990) Research into the functional hydraulic properties of geotextiles. *Proc IV Int. Conf. on Geotextiles, geomembranes & related products*, The Hague, Vol. 1, 285-288.

FISCHER G.R., CHRISTOPHER B.R. and HOLTZ R.D. (1990) Filter criteria based on pore size distribution. *Proc IV Int. Conf. on Geotextiles, geomembranes & related products*, The Hague, Vol. 1, 289-293.

GIROUD J.-P. (1982) Filter criteria for geotextiles. *Proc II Int. Conf. on Geotextiles*, Las Vegas, Vol. 1, 103-108.

GIROUD J-P. (1988) Review of geotextile filter criteria. *Proc. First Indian Geotextiles Conf. on Reinforced soil and geotextiles*, Indian Institute of Technology, Bombay, Vol.1, 1-6.

GOURC, J.-P. & FAURE, Y.-H. (1990) Soil particles, water and fibres - a fruitful interaction now controlled. *Proc IV Int. Conf. on Geotextiles, geomembranes & related products*, The Hague, Vol. 3, 949-971.

HALIBURTON T.A. and WOOD P. D. (1982) Evaluation of US Army Corps of Engineers gradient ratio test for geotextile performance. *Proc II Int. Conf. on Geotextiles*, Las Vegas, Vol. 1, 97-102.

HEERTEN G. (1986) Functional design of filters using geotextiles. *Proc III Int. Conf. on Geotextiles*, Vienna, Vol. 4, 1191-1196.

HOLTZ (1990) Discussion. *Proc IV Int. Conf. on Geotextiles, geomembranes & related products*, The Hague, Vol. 3, 1051-1052.

KOERNER R.M. and KO F.K. (1982) Laboratory studies on long-term drainage capability of geotextiles. *Proc II Int. Conf. on Geotextiles,* Las Vegas, Vol. 1, 91-96.

KOERNER G.R. and KOERNER R.M. (1990) The installation survivability of geotextiles and geogrids. *Proc IV Int. Conf. on Geotextiles, geomembranes & related products,* The Hague, Vol. 2, 597-602.

OGINK H.J.M. (1975) *Investigations on the hydraulic characteristics of synthetic fabrics.* Delft Hydraulics Laboratory, Publication No. 146.

RANKILOR P.R. (1981) *Membranes in ground engineering.* John Wiley, London, 377pp.

ROLLIN A.L., MLYNAREK J. and VIDOVIC A. (1990) Filtration opening size of geotextile - the parameters influencing the pore size determination. *Proc IV Int. Conf. on Geotextiles, geomembranes & related products,* The Hague, Vol. 1, 295-299.

WILLIAMS N.D. and ABOUZAKHM M.A. (1990) Evaluation of soil-geotextile filtration characteristics using the hydraulic conductivity ratio analysis. *Geotextiles & Geomembranes* Vol. 8, No. 1, 1-20.

WILLIAMS N.D. and LUETTICH S.M. (1990) Laboratory measurement of geotextiles filtration characteristics. *Proc IV Int. Conf. on Geotextiles, geomembranes & related products,* The Hague, Vol. 1, 273-278.

The way forward: a manufacturer's view

K. BEUDEKER, General Manager of Akzo Industrial Systems

The 25-30 years old history and origin of geosynthetics as spin-offs of existing industries: textiles, industrial fibres, non-wovens and wovens, are well known. In the future there will be even more companies involved with the quick diversification of products and applications, the growing variety of raw materials and composites.

During the first one or two decades non-woven geotextiles were emerging, representing 70 to 90 percent of the usage of geosynthetics. For the more demanding applications stronger woven fabrics were used as reinforcement, mattresses, bags, etc.

The geosynthetics market has been built up with a tremendous effort from the industry for product and application developments. The development has been a very complex and costly operation to prove, convince and make the civil engineers, landscaping architects, scientists and project owners aware of the safety, the economic and technical advantages of geosynthetics.

In general over the years industries have been making limited profits. The profits have not been sufficient to pay for all developments and the costs of marketing new products.

On one hand existing industries have been looking to the upcoming geosynthetics as a possible outlet for over-capacity. On the other hand the reduction demand for traditional textiles has resulted in a trend for companies to enter into the new fields for their products. Summarising the perceptions over the last decade of quite a number of industries active in Geosynthetics, the following general features can be identified.

(a) The volume of more than 50% of the business has been incremental to business in other fields.

(b) The profit margins (if any) on the low strength products, which make us a large part of the geosynthetic market, are too low to pay for stand-alone investments with an acceptable financial return on investment.

(c) A number of industries have subsided or are still subsidising their geosynthetic activities to keep their products in the market.

(d) The European geosynthetics industry has paid a 3 to 4 times higher marketing cost for products than the American industry. The higher costs are associated with making provision for all the different national requirements in Europe, language, legislation, etc.

There are several isolated successes where companies have managed to make good returns for products in the European geosynthetic market.

Consider the long procedure we have at the start of a project: all the discussions between designer, owner, specifier and manufacturer usually get a good design; then too often the specification is ignored on site for the sake of an alternative cheaper product, which is often untested and which has no historic record of performance. A quick test may be carried out. The manufacturer who assisted earlier in the project with development, long term testing and for a substantial investment loses the job to an interloper with an inferior product. The client then receives a substandard project, the quality may be low and the previous development costs will need to be recouped from elsewhere on another project.

Some approval procedures actually serve to restrict trade, good specification is a must but they must be fair, particularly in the European market. If we continue to act in the current way we will restrict market development.

We are also missing detailed on-site quality checking of geosynthetics delivered to site, this is allowing easier access into the market for cheap substandard products. Engineers and clients must insist and check that materials supplied do in fact meet their specification.

I am confident that the situation can be improved and the key points are

(a) development of realistic standards

(b) reliable testing procedures

(c) more uniformity in design criteria

(d) standardisation of product data sheets

(e) general introduction of quality control before geosynthetics are installed and checking of workmanship on site

(f) industry must participate in the quality control procedures

(g) well co-ordinated co-operation between all who are part of the geosynthetic community through IGS, EAGMID. We must not duplicate efforts and must all work together.

The future view is that the geosynthetic industry can survive and develop new products, and we would of course like to see a large expansion of the market to ensure the economies of scale.

The way forward for European harmonisation: assessment of the hydraulic characteristics and properties of geotextiles

B. MYLES, Chairman BSI TCM/35 (Geotextiles) and Convenor of CEN-189 WG.4 (Hydraulics), UK

1. Why is harmonisation needed?

The production and marketing of geosynthetics is an international operation, with major companies selling a range of similar products to different countries.

Civil engineering is traditionally a nationally based discipline with regulations and requirements specific to a geographical Area or Authority. These two factors, international production and national requirements, give rise to inconsistencies and difficulties, particularly in the area of test methods which are used to assess and characterise properties. Not only do these difficulties become a hindrance to trade and the free flow of goods but the national differences can be manipulated to bar certain products and favour others.

2. What properties are being considered for harmonisation?

The testing procedures for mechanical and hydraulic properties are being assessed as well as the criteria and conditions by which the durability of the geotextile will be judged. This note will deal only with the harmonisation of hydraulic testing of geotextiles and related products.

3. How is harmonisation taking place?

A technical committee (TC-189) has been established within the European Committee for Standardisation (CEN) to agree common testing procedures, methods of identification and assessment techniques, for geotextiles and related products. This committee has its secretariat in Belgium and is subdivided into five Working Groups:

WG.1 Direction and Co-ordination

Geotextiles in filtration and drainage. Thomas Telford, London, 1993

WG.2 Identification and Classification
WG.3 Mechanical Properties
WG.4 Hydraulic Properties
WG.5 Durability

The creation of this CEN committee has put a 'hold' on all new geotextile standards in the participating countries. The committee will allow standards which have been completed or in the throes of being printed to be published but all future standards work must be published as a 'Draft for Discussion' or a 'Discussion Document.

4. Who is involved in harmonisation?

All EEC and EFTA countries signed a Treaty in 1984 agreeing to participate in the co-ordination and harmonisation of standards, therefore the term 'the whole of Western Europe' can be said to be an accurate description. Many of the topics being considered by CEN-189 are being studied by the International Standards Organisation (ISO) Geotextile Committee (TC-38 SC.21). ISO representatives also take part in the CEN TC-189 discussions as non-voting members. Decisions are made, wherever possible, by consensus; narrow majority decisions are avoided, as the complex weighted voting system applied to the ratification of standards can block proposals which do not have overwhelming support.

5. What is the scope and aim of the harmonisation?

The present aim of CEN TC-189 is to produce a grouping of relevant test procedures that will be common to all participating countries and the presentation of index values or assessment levels will be the same.

6. Which hydraulic characteristics are being considered for harmonisation?

6.1. Water flow through the geotextile

This test procedure relates to uni-directional flow. Whilst there are variations in the arrangements of the test apparatus used, the main area of harmonisation will be in the water quality and presentation of results. Water flow as a series of hydraulic conditions is a requirement of the test in order to produce an envelope of hydraulic behaviour from which water flow and water head index are extracted. There will be a computerised falling head procedure appended to the main standard in which the hydraulic behaviour envelope and the two index values are directly computed.

6.2. *Initial water resistance*

The difficulty in appraising the water head needed to initiate flow through a geotextile has been a long-standing problem. The uniformity of water penetration over the surface of the geotextile is not easily obtained by standard water flow testing. Several countries supported the inclusion of this test procedure which is based on the French standard.

6.3. *Pore size*

There are three different pore size determination procedures in common use in Europe: Dry Sieving, Wet Sieving and Hydro-dynamic Sieving.

The task of harmonising the test procedures has proved difficult. The initial work was carried out in ISO TC-38 SC.21 and showed that, whilst there was little difference in the results on some geotextiles, there were significant differences on others. The strategy to achieve a harmonised pore size index is to have common particle and sample preparation, and common analysis and reporting of results; in addition the dry sieving method will be restricted to certain types of geotextile. Researchers in Holland, Belgium and Germany believe that this strategy will allow the different countries to keep their existing apparatus and basic procedures but achieve an indistinguishable pore size index.

6.4. *Water flow within the geotextile*

The uni-directional transmissivity testing procedure is very similar in most of the European countries. The main areas of harmonisation will be water quality, hydraulic gradient range and levels, compressive load range and levels together with compressive surface texture and composition. The analysis and reporting of the transmissivity values is also being harmonised. There is agreement amongst the countries that transmissivity values cannot be reported in isolation and that long term compressibility measurements of the geotextile or related product must be appended to the transmissivity values and that the compressibility behaviour must be with and without shear loading.

7. What changes will this harmonisation of hydraulic standards bring about in the UK?

Very little practical differences will be noticed in the UK. The main changes will be

(a) more information on the permeability of geotextiles, with the laminar and turbulent behaviour being better identified
(b) the initial water resistance value will be of interest but will fade as an issue when the manufacturers take remedial measures
(c) the introduction of hydraulic sieving for thicker and fine fibre geotextiles will lead to better and more reliable results
(d) the linking of transmissivity values with long term compressibility, particularly with inclined loads, will in some cases reduce the published transmissivity expectations of geotextiles and related products.

8. What form will the new standards have?

Each standard will be a stand alone document but may be part of a set of standards. The standard will be published in English, French and German with subsequent translation to other European languages. The CEN Standard will be adopted as a British Standard and be given a parallel BS Number.

9. What force will the harmonised standards have?

Not all the testing and assessment procedures being considered in CEN-189 will obtain a mandate from the Council but the three main hydraulic standards, Water flow, Pore size and Water flow within the plane, are certain to be mandated. This will mean that the European Standards will be mandatory for all public procurement projects and, as most of the projects in which geotextiles are used are financed wholly or partly by public funds, this will mean that in practice all geotextiles will be produced and sold to CEN Standards.

10. Will there be any further hydraulic standards for geotextiles and related products?

No new hydraulic tests are expected in the foreseeable future although there is some demand for hydraulic systems testing, i.e. geotextile-soil. The technical argument for this type of test is a strong one, however these test procedures, if agreed, are likely to be of an advisory nature rather than mandatory.

11. Will the harmonisation lead to a classification system?

Geotextile classification in some form will and must develop in time but this may relate initially to identification for trading categories or some other non-engineering reason. For geotextile applications the first category liable to classification is simple separators; nevertheless this is a contentious topic with strong views being expressed by several countries. Geotextiles for hydraulic applications are not likely to be considered for classification for quite some time, if ever, due to the wide range of parameters that are encompassed in a single product.

12. Will a common design procedure emerge from standard tests and classifications?

The 'dream' or 'nightmare' that one single design procedure will be applied Europe-wide to geotextile filters and drainage is just not true. There will be increasing convergence in design approach as theories are confirmed by practice and as experience and knowledge are exchanged. However, traditions, soil conditions and the availability of geotextiles are very different across Europe, and the CEN organisation and committees respect the autonomy and acknowledge the freedoms that designers of geotechnical hydraulic systems require.

13. Concluding remarks

This brief overview of the European harmonisation procedure for the hydraulic properties of geotextiles is a purely personal one; it does not represent the views of BSI TCM/35, CEN-189 WG.4 or any of my colleagues. The work of the CEN-189 Working Group 4 is still in progress and will take further time to complete its work and prepare the Standards for voting.

It would be wrong to present such an individual view without recognising the considerable work done prior to the establishment of CEN-189 by the BSI TCM/35 and ISO TC-38 SC.21 committees. I would like to thank the members of CEN-189 WG.4 for their outstanding efforts and co-operation, and in particular my long suffering UK hydraulic expert Paul Johnson.

The way forward: the client's view

P. JOHNSON, Highways Engineering Division, Department of Transport, UK

1. Introduction

The Secretary of State for Transport, as Highway Authority, maintains and keeps under review a national system of routes for through traffic in England. This is the Trunk Road system and includes Motorways. The purpose of the road network is to transport people and goods to their destination in a safe, economical and environmentally acceptable manner. The planning, design, construction and maintenance of highways is carried out by professional engineers who have a great number of factors, sometimes conflicting, to consider in carrying out their work.

The Department of Transport is often criticised for being slow to adopt new ideas and technology and being resistant to innovation and change. This is based on a misunderstanding of the Department's responsibilities and role. In fact the Department is very keen to explore new ideas and adopt new methods, materials and procedures provided these represent good value for money.

A key responsibility of the Department is the provision and maintenance of the transport infrastructure and not the development of new ideas or promotion of UK industry *per se*. This is reflected in the resources which the government allocates to research, construction and maintenance. However, within these constraints the Department is always prepared to support innovation to help achieve the best long-term value for money by encouraging appropriate new or improved technology. In road construction value for money often depends on obtaining adequate durability for both the component parts and the scheme overall and this is a topic which generates much discussion between innovators and the Department's engineers.

Geotextiles in filtration and drainage. Thomas Telford, London, 1993

2. Managing innovation within the department

The Department has a well-established procedure for the successful introduction of innovatory ideas, techniques or products. The present procedure involves the four stages of research, experimental trials, the development of specifications and standards and the longer term monitoring of performance in service. Not all of the stages will be required in every case but the general principles will normally be followed.

2.1. Research

The Engineering Policy Directorate of the Department currently spends about £10M each year on research on highways and structures in support of the development of specifications, standards and advice on good engineering practice. This can be set in the context of the whole road programme in which some £1395M will be spent on new construction in 1992-93 and £618M on maintenance (including structural and current maintenance) of the trunk road network. Some of this research is genuinely innovative, while some is the development and refining of other people's ideas or practice overseas. When research, which can include desk studies, laboratory testing, computer modelling or site testing, shows that a particular material, product or technique has potential for use on the Department's roads, full scale trials may need to be carried out on-site.

2.2. Site trials

The purpose of full scale site trials is to establish the performance in real site conditions, to investigate any installation or operational problems and to assess the likely cost and effectiveness under carefully controlled conditions. The experiences from the research and site trials are then combined in specifications, standards and advice notes.

2.3. Specifications and standards

Specifications are initially drafted for site trials and modified in the light of experience. Before specifications or standards are adopted by the Department they are subjected to rigorous scrutiny and the views of selected interested parties outside the Department may be sought. Departmental standards, where necessary, are developed in a similar way. It is the Department's policy to refer to established British Standards for products where appropriate, and in the future will refer to relevant European standards rather than developing its own. To that end the Department has made a significant contribution to the development of British and European standards, both in the committees and in supporting the necessary background work.

2.4. *Monitoring performance*

When an innovation is introduced its performance is monitored, firstly as part of the general monitoring of the network by the Department's Agents who are responsible for inspecting and maintaining the network. The Department's Headquarters also receives feedback directly from engineers in the Regional Offices and through a formal reporting system for defects known as the Quality Control Reporting System. In specific cases long term monitoring of innovations will be carried out for the Department by the Transport Research Laboratory or others.

In some cases ideas or techniques will already have been developed beyond the need for research and in such cases trial on site is normally required to ensure that any claims made can be substantiated in practice. However, for any proposed innovation the Department makes an assessment of the likely costs and benefits. In many cases the decision is made to continue with established practice because the benefits of the new technique are not readily apparent to the Department with its concern for safety, value for money and long term durability. Due to the costs of traffic management and gaining access the cost of failure of an unsuccessful innovation for the Department can extend far beyond the simple cost of repair or replacement. There may be even greater costs to the road user and the general public. In addition, should an accepted new product or material fail, say, three or four years after first use there may have been large quantities employed during that period all of which may need replacing.

3. The innovator's view and the client's view

The viewpoint of an inventor/manufacturer is often rather different from that of a prospective user, such as the Department. The innovator may have been familiar with, or at least aware of, the product or technique for five to ten years or longer and this will tend to give him confidence in the product. Human nature is such that the inventor and others involved in the development will normally, and understandably, have much enthusiasm for their ideas. In addition, there may well have been much investment of time, money and effort in developing the product or technique and it would seem only equitable that inventors, developers or entrepreneurs should be rewarded for their original ideas and effort. The patent system is intended to ensure fair recognition and reward for useful, original ideas. It is possible that these factors can sometimes lead to an over-enthusiastic presentation of the case in which the benefits are emphasised and the possible disadvantages are not recognised.

From the client's point of view the innovation is often intended to perform (with some advantage) the same function as an acceptable, existing system. Although a new product or technique may perform its specific function well, the client has to take the wider perspective of how the product relates to the overall scheme to ensure best value for money. Almost by definition an innovatory system will have greater uncertainties associated with it than a well known and well used technique. Whilst it is fairly straightforward to investigate the installation and short term performance aspects of new techniques, the medium and longer term performance are much more difficult to assess with certainty.

4. Barriers to innovation - apparent or real ?

In highway construction there is often a very long lead time in developing projects so that new ideas accepted by the Department may take several years before they are seen on the network. In addition there may be some element of conflict in a specification between using existing recognised standards (e.g. products Kitemarked to a British Standard) to give a simple method of ensuring fitness for purpose and having a more flexible approach which encourages new techniques and products.

The Department is aware that even when it has accepted an innovative technique for design or construction there may still be a slow take-up on road schemes. It has been suggested that there may be disadvantages in the UK's traditional construction system of client, design organisation, Engineer and contractor, one of which may be the lack of encouragement of innovation.

The Department is currently investigating Design and Build contracts and a number of trial schemes are underway. In the Design and Build approach the client normally states his requirements more in terms of performance, and the method of achieving that performance is largely chosen by the contracting/designing consortium. The benefits and risks are distributed differently from the current system and innovations offering financial savings overall,which meet the performance requirements, may be more readily adopted. The Department will assess the performance of these new contracts and, if satisfactory, they may have a place in the Department's future contracts strategy

5. Geotextiles and geotextile related products

Geosynthetics are a relatively new development, perhaps rivalling steel or concrete in their range of applications in civil engineering. Standards and designs using the more traditional materials have evolved over many years and people involved with geotextiles are faced with producing information, increasing confidence and establishing accepted practices in a much shorter timescale. Geotextiles can perform several functions and one of the Working Groups on the CEN Geotextiles Technical Committee (CEN/TC 189) has identified separation, filtration, drainage, protection, reinforcement and containment as individual functions. More than one function will be involved in most applications and the CEN Committee are trying to develop a classification system in which functions are categorised as 'primary, important or secondary' for each application.

Professor Koerner, director of the Geosynthetics Research Institute at Drexel University, made some very useful points during the Mercer Lecture at the Institution of Civil Engineers earlier this year. His lecture was titled 'Remaining technical barriers to obtaining general acceptance of geosynthetics'. After speaking in some detail about the technical obstacles, he reviewed four non-technical issues that he considered to be barriers to the wider adoption of geosynthetics. These were

(a) lack of formal instruction
(b) unfamiliarity with design and testing
(c) relative newness of the technology
(d) hearsay stories about degradation.

Professor Koerner offered a number of suggestions about how to overcome these barriers which generally involved ensuring that information and experience on the use of geosynthetics was made readily available to engineers.

6. The department's specification

Some of the uses of geotextiles in Department of Transport schemes are covered in the recently published 7th Edition of the Specification for Highway Works (SHW) and the 4th Edition of the Highway Construction Details (HCD). These and other similar documents are now produced in loose leaf format to ease the introduction of amendments and additions, which may be made on an annual basis.

Reinforced or anchored earth structures need to comply with Clause 2502 of the specification and proprietary systems require a British Board of Agrément, Roads and Bridges Certificate.

Geotextiles used for separation are covered in Clause 609 of the SHW and the corresponding Note for Guidance. The Department is looking to revise and expand the advice on geotextiles used for separation as experience develops. Advice is currently being developed to give guidance on drainage and separation layers using natural materials and geotextiles both alone and together.

Clauses 514 and 515 of the specification and also HCD give the requirements for fin and narrow filter drains for use as edge of carriageway drains. The Department has set up a testing regime in conjunction with the British

Board of Agrement leading to a BBA Roads and Bridges Certificate for such products and this certification will be required in future for use of such products on Departmental schemes.

The geotextile must be tested and certain values of tensile strength, puncture resistance and tear resistance met in order to obtain the BBA certificate. Water flow normal to the plane of the geotextile and the apparent pore opening size, termed O_{90}, must be measured and quoted on the certificate. For fin drains the transmissivity of the composite drain must be measured for water travelling approximately vertically downwards to the pipe for Types 6 and 7 and also down a gradient of 1:10 for the Type 5 where the body of the fin carries out the function of the carrier pipe. The Department has developed a system which attempts to simulate the possible long-term thickness of the installed fin drain and the transmissivity is measured in this simulated long-term (11 year) condition.

There are no 'pass or fail' values for the water flow, apparent pore opening size or transmissivity, as the requirements will be site-specific and the Engineer will have to check the values on the BBA certificate against his design requirements. It is possible that at some future date values or bands could be specified in some classification system but it is felt that this may still be some distance away. Geosynthetics are used in a number of other applications on highway schemes. Where the use is not covered by the Department's documents then the Design Organisation is responsible for producing that part of the specification, in consultation with the Department if necessary. Wick drains are quite often employed to speed consolidation of compressible ground beneath embankments. Because they are required to function for a relatively short period, typically two years, they are more readily accepted by both designer and client.

Geotextiles may also be employed on a scheme at the Contractor's instigation where he feels that time or other savings make it worthwhile. Such use would be subject to the approval of the Resident Engineer and/or the Department.

7. The European dimension

The spirit of the Treaty of Rome and the Single European Act, as expounded in the Public Procurement Directives and other documents, will become more evident as time goes by. Public bodies such as the Department are required to ensure that their procurement systems contain no barriers to trade within the community. This means that the emphasis is on accepting any goods and services which are fit for purpose. Purchasers will use Harmonised European Standards or European Technical Approvals (generally supervised by the Agrement Boards) where possible, but where such standards are not available the situation is rather more complicated. Purchasers may specify products as complying with a national standard or equivalent, and where suppliers offer products complying with equivalent standards of fitness for purpose such products must be considered. The major difficulty at present is how to judge equivalence between standards and who is to make the judgement.

8. Conclusions

The Department is supportive of innovation in design, materials, products and techniques within the resources available, but will want to be satisfied on value for money and safety before introducing them. It will always encourage competition between new and established materials, products and techniques.

The Department is investigating Design and Build contracts to see if they offer advantages over the present type of contract.

The 7th edition of the SHW provides the opportunity to update the Department's specification at regular intervals to take full advantage of the rapidly developing technology in the field of geotextiles.

As a result of European harmonisation in general, the public procurement directives and the construction products directive, the Department will increasingly use European standards for design and specification.

In the interim, whilst European standards are being developed the Department will look to independent certification such as BBA certificates and quality assurance to ensure satisfactory performance from manufactured products such as geotextiles for filtration and drainage.

9. Acknowledgement

This Paper is published by permission of the Director General Highways of the Department of Transport.

Discussion

Reporter: C. Lawson, Exxon Chemical Geopolymers

K. ROWE, University of Western Ontario
In Table 1 Dr Ingold gave several values for O_{90}; what test method was used to get these values?

T. INGOLD, Consultant
O_{90} values were measured by wet sieving.

K. ROWE, University of Western Ontario
How much confidence do you have in the predicted values in Table 1 which vary by a factor of 10?

T. INGOLD, Consultant
Go to Table 1 and take out the 110 µm value; we then have a factor of 5 between values.

K. ROWE, University of Western Ontario
As a Consulting Engineer which design method do you favour?

T. INGOLD, Consultant
I tend to vary the method to suit the situation of the project, with the knowledge that none are 100% certain for any situation.

N. FRENCH, MMG
A great deal of effort and time goes into the preparation of specifications. I would like to see Engineers exercise better control over the implementation of specifications to avoid the often seen supply of substandard materials.

Geotextiles in filtration and drainage. Thomas Telford, London, 1993

C. LAWSON, Exxon Chemical Geopolymers
A general question to all speakers: we have all heard a lot about specification and instances where one product is pipped at the post by a rival. A lot of speakers have knocked the product-specific 'BRAND X' type of specification; at least with a specification you know what will be supplied.

S. CORBET, G. Maunsell & Partners
I will not agree with you and say 'yes, specify' because I believe that a performance specification which follows from classification in turn leads to a fair market for all producers. Unfortunately, there are rules set by public clients requiring engineers to specify fairly and not be restrictive by naming products. To specify Brand X and no other is not allowed unless there is only one product for that job. We must get away from Brand X specifications. If Brand X specifications are used they are very restrictive and can indicate that the designer/specifier have simply used a product because it 'worked down the road'.

A further point about design: we have seen today that there are numerous different design rules in use. In my opinion Eurocode EC7 should include design rules for filters. I have not seen any indication or draft for such rules in the Eurocode EC7. We need a basic framework of rules into which the national bodies can put their national factors or prescriptions for design.

Can we send from the geosynthetic industry a message to the drafting panel of EC7 that the code should include design rules for geosynthetic filters and drains? Reinforcements are dealt with in the UK in the draft BS 8006 but no equivalent exists for filters and drains.

D. MANN, Applied Geology
I can accept 'Brand X' specification but I do not believe it is possible to assess 'or equivalent' because all manufacturers produce data sheets with test data from different methods of test? Do we really believe that needle punched, a woven or heat bonded geotextiles will behave in the same way? I personally do not believe they will.

S. CORBET, G. Maunsell & Partners
If you specify Brand X, there is no equivalent; the Engineer's Representative on site will have no choice, and the words will be 'or equivalent' are meaningless. Unless the man on site has all the design details to hand he cannot approve anything but Brand X.

K. BEUDEKER, Akzo Industrial Fabrics
Is this Brand X or Brand Exxon?

C. TUXFORD, EEE Ltd
I take issue with Mr Corbet. On several occasions I have offered alternatives to named 'Brand X' products; the Engineer's Representative has referred the decision back to the designer who often has no idea why he chose 'Brand X' or similar. A problem often occurs in that, yes we will accept the alternative, but we need 1 or 2 years to consider the alternative!

N. FRENCH, MMG
We have seen many instances of mixtures of product-specific parameters in specifications which makes the Contractor's life difficult in selecting a geotextile.

Chairman's concluding remarks

It seems to me, as to Steve Corbet, that limit state design and objective criteria are the way of the future for geosynthetics. Having heard the papers I am now not as nervous as some seem to be about the mechanisms involved in filtering and clogging. We understand which piece of physics we are talking about, we simply have not agreed yet on exactly which tests to do and how to interpret the data for design. We are not casting around in the dark, these are materials and techniques which have been used for decades in some cases, and people have executed works, dug them up several years later and reported to us on the qualified and unqualified success that these techniques have given. Therefore I am not so pessimistic as some of you that we have to throw everything aside, saying 'I once used this class of material from that manufacturer, and as I don't really understand the issues I will only specify that material in future'.

I take comfort from the work that is going on with the standardisation of tests with CEN. We must be sufficiently professional at least in Europe to decide on the methods we are going to use. So long as we have a workable test which we all do in the same manner, it does not matter that it is not a particular individual's favourite test, provided we all follow the same procedure.

One aspect which has been discussed during the day is that all the good design and specification can be spoilt by what goes on on site. No-one has spoken about the importance of training technicians or technician engineers who may be carrying out the works, but I suggest that it behoves IGS to sponsor or encourage training courses for technicians. Quality Assurance Statements would then need to state that people with an appropriate qualification shall be present on site controlling the operations. Otherwise how can we have a quality product, if there is not proper control on site? Likewise with the education of Chartered Engineers, there must be a way to introduce

Geotextiles in filtration and drainage. Thomas Telford, London, 1993

geotextiles into undergraduate courses. One must understand how overcrowded traditional 3 year courses are; many universities are extending courses to 4 years. At Cambridge we are planning to include more about the special geotechnical process and design (such as geosynthetics) in the fourth year. IGS will still need to run special 1 day conferences such as Geofad '92 to ensure that all Engineers receive proper post-graduate development.

Other aspects of the way the industry is moving, particularly 'design and build' contracts, may give increased opportunities for 'high-tech' developments in ground engineering. Owners will also need to be educated that quality must be specified and then assured. Mr Beudeker said that many owners and government bodies who are effectively the owners of roads, etc., need to understand that engineers and manufacturers expect to work at a high professional level to deliver an economical product.

On the scientific side it is difficult to do fundamental research; we have heard that TRL choose to spend £10M on geotechnics out of a total annual budget of £200M on research in transport. This is a significant sum, but quite small in relation to the £2000M spent on road building and maintenance. The SERC spends about £1M per year on fundamental research in geotechnics, which is clearly inadequate. More must be known regarding the transport of individual particles and the effect of significant variables such as the local magnitude of effective stress, not only in the soil but also in the neighbouring geotextile filter. Perhaps the European Community can help: the EC research budget is larger and we must join with our colleagues in the other EC countries to provide research and development in this field.

The final points about science have emerged strongly during the day. People have been brave enough to do laboratory work even with landfill leachate; my heart goes out to everybody doing that work! Significantly, field studies have been reported, case studies, failures, successes. Surely through the reporting of field studies and laboratory work the state of the art will move forward - in 10 years' time it will seem inconceivable that anyone wanted to specify Brand X.

DURHAM UNIVERSITY LIBRARY
3 0104 00729135 8